ENGINEERS AND INDUSTRIAL GROWTH

Engineers and Industrial Growth

Higher Technical Education and the Engineering Profession During the Nineteenth and Early Twentieth Centuries: France, Germany, Sweden and England

Göran Ahlström

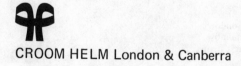

CROOM HELM London & Canberra

© 1982 Göran Ahlström
Croom Helm Ltd, 2-10 St John's Road, London SW11

British Library Cataloguing in Publication Data

Ahlström, Göran
  Engineers and industrial growth.
  1. Engineers—Europe, Western—History
  2. Europe, Western—Industry—History
  I. Title
  338'.06'094   HD2351

  ISBN 0-7099-0506-8

Typeset by Jayell Typesetting, London
Printed and bound in Great Britain by
Biddles Ltd, Guildford and King's Lynn

CONTENTS

TABLES AND FIGURE

Tables

Figure

TO THE Ms OF MY FAMILY

PREFACE

This study is partly the result of my work as a consultant to the British Department of Industry (DoI) in their *Engineering Professions Project* 1975-7. After my return to the Department of Economic History at Lund University, funds from the Swedish Council for Research in the Humanities and Social Sciences made it possible for me to complete the book.

I wish to express my gratitude to several people for discussions on the topic and for their critical comments on my reports to the DoI and/or on a previous article on the subject (*Economy and History*, Vol XXI:2, 1978): Sir Ken Berrill and Mr Michael Fores, Drs H.E.S. Fisher and Torsten Althin and Professors J.F. Baker, Sune Carlson, Erik Dahmén, Wolfram Fischer, Charles P. Kindleberger, Roger Kling, David S. Landes, Walter Minchinton, Frank R. Pfetsch, Nils Runeby and Rolf Torstendahl. Professor Maurice Levy-Leboyer kindly allowed me to use unpublished material on the French engineers.

I am especially grateful for the critical observations and suggestions offered by some of my colleagues and friends at the Department of Economic History on having read the manuscript: Professor Lennart Jörberg and Drs Olle Krantz, Björn Lárusson, Rolf Ohlsson and Lennart Schön. My warm thanks to them all.

A special thanks to Mrs Kerstin Bengtsson, who typed the manuscript, and Dr Marianne Thormählen, who scrutinised my English.

1 SCOPE OF THE STUDY

1.1 Introduction

In the middle of the nineteenth century – when the industrialisation of France had been under way for a few decades and Germany was still in an earlier stage of industrialisation – about 300 to 325 students graduated annually from the German *Technische Hochschulen*, thus named from the 1860s, while the figure for corresponding French schools was slightly lower. Around 1900 the number of annual graduates in France was approximately 1 000, while the number at the German Hochschulen was now three to four times greater than in France. In Sweden – where the process of industrialisation[1] lagged not only the French but also the German one – the annual number of graduated engineers from the corresponding two technical schools increased from about 30 in the 1850s to an average figure of about 150 in the years between 1910 and 1920.

Calculations of the total number of qualified engineers in the societies during the nineteenth and early twentieth centuries show that in the middle of the nineteenth century there were no more than 3 500 qualified engineers in Germany, while the number in France was twice as large.

On the other hand, by the time of World War I, Germany possessed a considerably larger number of qualified engineers than France, in absolute number; Germany then had 60 000 highly qualified engineers whereas the figure for France was slightly over 40 000. Still, regarded as a proportion of the economically active male population these engineers made up about 3 per thousand in Germany as well as in France around 1910.[2] In the case of Sweden, the total number of engineers graduated from the technical universities in Stockholm and Gothenburg was about 600 in the early 1850s – to which can be added a number of highly qualified military engineers slightly exceeding this figure – and about 2 000 at the turn of the century.

At the time of the war, the latter figure had increased by more than 50 per cent. In relative terms, the density of qualified engineers in Sweden in the early twentieth century amounted to two-thirds of the French and German figure. The situation in these three countries differed significantly from the case of England, where the annual number of graduates was very low. Around the turn of the century, the

13

total number of members in the four leading English engineering organisations – the Institutions of Mechanical, Civil, Electrical, and Mining Engineers – merely amounted to about 15 000,[3] which entails a relative proportion of only 1 qualified engineer per thousand of the economically active male population. Compared to the situation in the other countries in the beginning of the twentieth century, that is 50 per cent of the Swedish figure and only one-third of the French and German one. Furthermore, it should be emphasised that with regard to France, Germany and Sweden, the figures are based on the distinguished highly qualified engineers; where England is concerned there are, as will be indicated below, reasons for doubting the quality of the English education – and the English qualified engineer belonging to any one of the organisations mentioned above – when making an international comparison.

From the beginning of the 1830s up to World War I France had, according to the industrial production statistics, an overall growth of less than 2 per cent per year, while the corresponding figure for the United Kingdom was almost 2.5 per year.[4] If the whole period 1830-1913 is broken down into sub-periods of two to three decades, the growth in France only exceeds that of England during a period extending from the mid-1890s; at that time the French figure was 3 per cent a year, to be compared with the overall UK figure.

On the other hand, from the 1850s and 1860s respectively, Germany and Sweden experienced industrial growth rates that were among the highest in the world, an average of slightly below 4 per cent a year. Furthermore, in these two countries the long-term industrial growth rate increased from the beginning of the 1880s.

The basic question to be raised here is this: What connections can be found between these – and other – pieces of evidence concerning the highly qualified engineers in each country on the one hand, and that country's industrial performance on the other? It should be noted that the present study does not aim to supply *the* explanatory factor in the growth process; consequently, I have no faith in monocausal explanations in economic history. However, the quantitative evidence found here combined with information derived from the scholarly literature and other sources – quantitative as well as qualitative – indicate the great importance of the highly qualified engineer in the industrial growth process.

The present work focuses upon engineers educated at technical schools comparable to universities, who are *a priori* assumed to be of great importance for the development of industrial production. Thus

I accept a view which prevailed in France, Germany[5] and Sweden during the nineteenth century, and to which, from the middle of the nineteenth century, leading critics of the English engineering education also subscribed.[6] In order to understand, apply and further develop the latest knowledge in science and technology, an education at an advanced — for its time, that is — technical school was considered the best means.

The outcome of this view on the Continent is shown in the expansion of the countries' system of technical education. According to Landes, the French and German institutions particularly constituted 'a veritable hierarchy', with the schools discussed here on the highest level.[7] In Sweden a similar hierarchical system of the technical education, with higher and lower schools, was built up.[8] The same principles were also predominant during the twentieth century; according to, for example, a Swedish public investigation concerning the higher technical education (1943), it was emphasised that the goal of the technical universities was to educate those technicians 'who are meant to lead and develop the technical and commercial life'.[9]

Thus, the idea of the division of labour and its corresponding demands for technical qualifications in people employed in various occupations was emphasised from an early stage. Within industry, three main groups of tasks and levels were distinguished which required a technical education: the jobs that demanded a technical university education (or similar), those that needed an engineering education on a lower level, and those that only called for a limited technical education. The national system of technical education should provide the manpower for all these tasks and levels. The need for highly qualified engineers was thus realised at a stage in the industrial development when the science-based industries gained a leading position in the process of industrial growth. Consequently, a thorough knowledge of scientific principles grew increasingly important as a basis for industrial technology.

However, our knowledge of, for instance, the number of qualified and professionally active engineers in different countries during different periods of time is very limited. The educational and demographic statistics are highly deficient in this respect. An important object of economic-historical research will therefore be to fill in the blanks of statistical information through calculations and estimations, enabling us to discuss the industrial development of the countries from partly new points of departure.

The quality of the system of technical education in England, France and Germany has indeed been debated, but there has been no explicit discussion concerning, for instance, the extent of this system on a comparative basis. Apart from monographs on the Swedish technical universities, it is only recently that scholarly writers have begun to consider the Swedish development from certain educational and social aspects.[10]

It might sound trite to say that technical education has always existed in one way or another. The knowledge of how to make things has always been taught; father to son, man to man, master to apprentice. However, it is only when technical knowledge was brought into institutional forms that we can speak of a formal system of technical education. It was towards the end of the eighteenth century that this type of education was developed; first in France, then in the Austrian double monarchy and the German states. In England, on the other hand, the idea of 'learning by doing' was predominant, along with a long-lasting belief in the 'practical man'.[11]

Stating that the engineer has, in a wide sense, been an important factor during and since the industrial revolution in England (for the sake of simplicity, we retain this expression) in the emergence and development of the production apparatus of the industrial world may also seem to be a commonplace. It is clear, though, that the education of engineers as well as the social and legal status of the engineering profession varied in different countries. As a consequence of, among other things, the industrialisation process which started later compared to England[12] and different educational-philosophical traditions, technical educational institutions, which during the latter half of the nineteenth century were accorded university status, soon emerged on the Continent. Consequently, when these countries were industrialised they possessed corps of engineers with broad theoretical and practical knowledge in supervisory functions,[13] something that was generally lacking in England for a long period.[14]

In his study 'Technological Progress and the Industrial Revolution'[15] based on a survey of the existing literature, Lilley has asserted that in a technological sense, the industrial revolution in its early stages entailed an enormous acceleration of a process which had been going on since the early Middle Ages, and that the inventions were *primarily* made in response to economic demands; 'The early stages of the Industrial Revolution — roughly up to 1800 — were based very largely on using medieval techniques and on extending these to their limits.'[16]

As regards the cotton industry, which first exhibited the possibilities of a hitherto unheard-of expansion, Lilley states that the inventions made were ' "easy" . . . in the sense that they required no special qualifications or training'.[17] In the case of the iron industry, prior to the 1820s the only sector besides the cotton industry which had a 'highly successful, rapidly diffused technical change',[18] improvements in the quality of iron certainly implied solutions to difficult technical problems; still, according to Lilley, they involved 'no major discovery – merely gradual improvements in operating conditions, arrived at by patient trial and error.'[19]

It should be emphasised, however, that science was of great importance at an early stage, above all in the field of industrial chemistry; according to Lilley, science 'at the highest level then available'[20] was the key to technological innovation. Progressive manufacturers, particularly from the latter decades of the eighteenth century, were well aware of the importance of science. Although scientists were not directly employed, attempts were made to arrange for a gentleman with scientific leanings to live near the factory if he chose, thereby acting as a consultant in matters of science.

Thus, Lilley establishes that science during the eighteenth century and the beginning of the nineteenth century did not initiate industrial trends; 'It merely solved some problems which the general movement of industry had created . . . If persistence, ingenuity and craftsmanship could speed up the process or find the substitute as required, that was good enough. If they could not, then as a last resort science was called in.'[21] It is clear, though, that 'where science *was* used, it played an absolutely essential role'.

As the importance of science within industry increased substantially during the nineteenth century, the way in which theoretical knowledge in science was combined with practical experience in engineering also came to be of growing importance. It is in this respect that England has, since the beginning of the 1800s, generally deviated from the Continental development.

Nevertheless, it should be made clear that during the earlier industrialisation in the pioneering country, England, a technical education based on the principle 'learning by doing', empiricism playing a vital part, was adequate.[22] But with the growing importance of science and its application in industry, this type of education became insufficient. An adequate education for persons in leading positions in industry now meant – in England as well as on the Continent – a qualified theoretical and practical education based on scientific principles. Apart from certain criticism which, however, was mainly directed towards the position of basic sciences in England,[23] this fact did not become apparent to the English until the middle of the nineteenth century. Nevertheless, few changes were made after the realisation had begun to sink in.

Of course it is very difficult – perhaps impossible – to *measure* the importance of a specific type of engineer in the industrial growth process, but a comparative analysis of different countries concerning respective numbers of available engineers, and the functions that the engineers fulfilled, leads to indicative conclusions about their importance.

The relevant starting-point in such an analysis must be the number, in each country, of engineers with the kind of education discussed in this study, including among other things, a discussion of the demand for and careers of the engineers. International comparisons based on the national results will come next, similarities and deviating features being recognised and analysed.

1.2 Method

As our knowledge of the number of highly qualified engineers in the countries is small, the first aim of the study is to present calculations on the supply and the cumulated total numbers of these engineers during the nineteenth and twentieth centuries. Furthermore, by way of discussing the industrial demand for qualified engineers and the engineers' choice of professional categories after completed education, as well as by way of studying the social background of the engineers, I shall consider the views on the social status of the engineering profession and the higher technical education in the countries. The social standing of a profession and a specific education is an important aspect, as it tells us a lot about technology as a cultural factor in a society. Although economic motives are of course important when choosing a career, I assume that there is a positive correlation between the choice of a career and the status of a particular occupation. A high-status occupation attracts well-qualified manpower. Similarly, if a particular education has a high social standing, it attracts the most able students when they choose their further education after finishing secondary school.

The discussion – which is based on the evidence supplied by my calculations – leads to conclusions that might have been specified in terms of 'lack' or 'shortage' of highly qualified engineers within industry in certain countries. From a purely economic point of view, however, it is incorrect to talk about 'lack' or 'shortage' if the demand for this type of manpower was also small. Following an expression coined by J. Jewkes, it is more appropriate to use the term 'unmet need' in such cases, a term implying that not merely the supply but the demand, too, was inadequate.[24] Jewkes discusses four ways of identifying a situation of an 'unmet need': (1) international comparisons; (2) industrial comparisons; (3) establishment of positive correlations between industrial output and number of scientists and engineers employed in industry – on the presumption that an increase in the number of these personal categories will increase industrial production; and (4) measurement of net return for specific inventions or particular industries or countries, and of investment in research and development.

The present study is in line with the first and third approaches. Certain features in the development of the higher technical education in the countries concerned will be emphasised, too, and in that context the role of the state will also be considered. The calculations and arguments presented lead to conclusions about the importance of highly qualified engineers for the industrial development of the four countries during the studied time.

As England stands out as the exception from the general pattern of the institutionalised higher technical education – as well as concerning the process of professionalisation of the engineers – it is reasonable to treat the English development in a separate section.

Since the national statistical data concerning the engineers are sadly deficient, this study has, by and large, had to rely on the data for individual schools.

Still, aggregate figures covering the number of students at the German Technische Hochschulen during different years, as well as aggregate figures pertaining to the corresponding two Swedish schools, are relatively easily accessible;[25] but as regards the higher technical schools in France, no such aggregate information is available prior to 1970.[26]

For a discussion of the students' field of specialisation and occupational choice after completed education, their social backgrounds, etc., we have to depend on material and studies concerning individual schools in all countries.

Here, I have selected what might be termed 'model schools' with regard to the shaping of the higher technical education and schools with a good reputation; as far as the status aspect is concerned, these are thus the 'distinguished' engineering schools. The material indicates that these aspects are compatible.

For reasons stated below, the following institutions have been chosen for the study: *École Polytechnique, École des Mines* (Paris) and *École Centrale des Arts et Manufactures* in France, the *Technische Hochschulen* in Karlsruhe and Berlin in Germany and, for Sweden, the technical universities in Stockholm and Gothenburg.

Finally, three limitations in my approach should be mentioned. First, I have had to make the study at an aggregate level only. Its scope has not allowed for any comparative micro-level investigations between the countries concerning various industrial branches.

Secondly, I only deal with the highly qualified engineers; consequently, the engineers with lower technical qualifications are not taken into account. However, it may well be assumed that my conclusions would have been strengthened if these engineers had been included in the calculations and discussions, too, for example those with an education from the technical secondary schools and technical institutes. In Germany at the turn of the century, for example, these engineers have been estimated to have been three to five times the number of those graduated from the technical universities[27] and in the case of Sweden, the figure covering the highly qualified engineers will have to be multiplied one-and-a-half times in order for us to obtain the total number of technicians in Sweden at the beginning of the twentieth century.[28]

On the other hand, in the case of France, the number of engineers with lower technical qualifications was obviously more limited. Kindleberger has made this point while discussing various criticisms of the French scientific and technical education, especially at the *Grandes Écoles*. According to the most important criticism, the educational system and its output concentrated too heavily on the top ranks of instruction, administration and industry, and neglected to provide sufficient training for instructors, laboratory assistants, middle cadres, foremen and skilled workers. Although successive attempts were made to fill the gap with various types of schools, Kindleberger could establish, 'As in underdeveloped countries today, France in the nineteenth century had too high a proportion of highly trained theorists relative to lower ranks who would carry out their directions in production.'[29]

Where England is concerned, the engineers who had received their training at the technical colleges, institutes and similar schools are included in the figures stating the number of members in the engineering organisations mentioned above. Unfortunately we do not know the relative proportions of the various types of educational background that the members had received; nor can we establish to what extent the engineers were members of an engineering organisation. However, there is good reason to assume that a very large proportion of the English engineers – perhaps more than 85 per cent at the turn of the century – were members of an organisation.[30]

The essential feature here is thus the comparatively small number of engineers in England, even when those with lower technical qualifications are taken into account. Also, certain characteristics of the French development brought up and discussed in this study receive further confirmation.

Thirdly, a minor limitation in the calculations stems from the fact that a small proportion of the students at the technical universities were foreigners,[31] as well as the fact that a certain number of the practising engineers in a country had trained abroad.[32]

1.3 Definitions

While it is fairly easy to determine the contents of 'higher technical education' – that is, education provided by those schools which were already, or were later during the nineteenth century converted into, technical universities – the following remarks should be made concerning the concepts of 'engineering' and 'profession'.

It is impossible to give an exact definition of 'engineering' which would be valid both in space and time. It is clear, though, that 'engineering' as an 'art' existed long before engineering as a 'profession'[33] and that in the beginning of the nineteenth century it was only in France that engineering was 'clearly and definitely established as a learned profession'.[34] The term 'engineer' emerged during the Renaissance when the needs of war stimulated the development of 'engines of battle';[35] as late as 1805 in Sweden, an 'engineer' was defined as 'a military man who understands fortification in the field as well as in the fortress'.[36]

At first, then, engineering was developed as an occupation within the military sector; in order to distinguish the non-military engineers, they were termed civil engineers.[37]

In the foreword to his study *Engineering Education: A Social History* Emmerson writes: 'Modern scientific principle has been drawn from the investigation of natural laws, technology has developed from the experience of doing, and the two have been combined by means of mathematical system to form what we call engineering.'[38]

Such a definition would at first appear to be simple and lucid, but there are two reasons why it cannot be considered quite satisfactory. In the first place, 'the experience of doing' is not an exhaustive definition of the technological element in engineering. Second, it is a dubious claim that 'technology' developed solely on the basis of practical experience.[39]

The term 'technology' means 'the knowledge of "technique" ' and although it implies the practical arts in wider sense, technologies are 'bodies of skills, knowledge and procedures for making, using, and doing useful things'.[40]

Hence, engineering is − like 'technology' which should, according to Emmerson, help define 'engineering' − a broad concept which lacks a specific content for a certain period of time and whose connotations keep changing.[41]

Instead of attempting to supply a precise definition of engineering and technology, it is more essential to stress that engineering constitutes a part of the concept of technology. Another important point is the fact that the philosophy of technical education underlying the engineering activities and the education of engineers varied in different countries.

As a consequence of this, among other things, the process of professionalisation of the engineering activities seems to have varied as well. The core in the theory developed around the professionalisation is the distinction between occupation and 'profession', which implies that the occupational categories are considerably more than occupations and that an occupational category can, with the help of certain criteria, be developed into a profession, i.e., be 'professionalised'.[42]

What criteria to apply, and what method of analysis is preferable, are sophisticated sociological problems which cannot be considered in this context. As regards the choice of the method of analysis, however, it seems reasonable to choose a gradualistic method − in contrast to a typological one − in which all the occupations are considered to be more or less professionalised. 'What is or is not a profession is determined by the intersection of the scale of professionalisation at some point, the occupations above this point being denoted as "professions" '.[43] A profession is hence briefly defined as an occupation whose members possess a high degree of *specialised, theoretical knowledge*, are expected to carry out their tasks while taking certain *ethical rules* into account, and are held together by a strong sense of *esprit de corps* arising from a common education and adherence to certain doctrines and methods.[44]

As we have seen, it was only in France that engineering could be considered as a learned profession in the beginning of the nineteenth century. This was above all due to the position of technical education in France;[45] but it was also a consequence of the fact that the other criteria presented above could be said to have been fulfilled.

The national organisation *la Société des Ingenieurs civils de France* was founded in 1848 by previous students of the École Centrale des Arts et Manufactures, a private institution at that time.[46]

In the case of the German states, there was a rapid development during the first half of the nineteenth century through the *Gewerbe* institutes, founded from the 1820s onwards. Former students at these institutes founded associations. A reorganisation in 1856 of the association formed ten years earlier by the former students of the Berlin Gewerbe institute also resulted in a national organisation, the Society of German Engineers (*Verein Deutscher Ingenieure, VDI*). From the start, the *VDI* took considerable interest in educational questions, and the focal point of the *VDI*'s interest ever since its foundation was the rank and position of higher technical education.[47]

Thus it can be asserted that the process of professionalisation of these engineers was far advanced in the middle of the nineteenth century by means of, among other things, the knowledge acquired at the polytechnical schools and the *esprits de corps* that prevailed. In the middle of the 1840s, it was also claimed that 'higher' technicians and *Maschinenbaumeister* belonged to a special class together with the theoretically *Gelehrten*.[48] Besides the *Gebildeten* and *Studierten* within the humanities and the mathematical-natural sciences, there was now a technical-industrial field, too. A potent self-confidence (*Selbstbewusstsein*) could thus be observed among the technical-scientifical interests in Germany, where the leading stratum comprised the polytechnical schools and the engineers educated at these schools.

A similar development — but with a certain lag — can be seen in Sweden, where the Swedish Association of Engineers and Architects (*Svenska Teknologföreningen, STF*) developed out of a student organisation at the technical university of Stockholm. The association considers 1861 as its year of foundation, but it had forerunners in the 1850s.[49]

The Stockholm *Ingenjörsföreningen*, founded in 1865, should also be mentioned here. It emerged as an elite group, including persons who were considered the most prominent of Sweden's technicians.[50]

This organisation was also in close touch with scientific research and technical education. Of its total number of members in 1865 (95 persons), about 80 per cent had a theoretical education of some sort, and one quarter were graduates of the technical universities of Stockholm or Gothenburg.[51]

In 1891 Ingenjörsföreningen was amalgamated with Svenska Teknologföreningen. That further increased the importance and strength of this organisation, which formed a national association from the latter half of the 1880s onwards. It became 'a point of support for the efforts of engineers to win recognition, on the national as well as the local level, for technical and industrial points of view . . .'.[52]

However, it should be emphasised that the Swedish process of professionalisation was well developed as early as 1870 and that the activities within the two important organisations were not limited to pure technical problems. This has been pointed out by Runeby in his work on the Swedish engineers; Runeby also states that this new elite emanated from science. Secure in the knowledge of his own competence, the highly qualified engineer also claimed his place – and responsibility – in the society.[53]

As was indicated above, and will be discussed further in Chapter 3, the process of professionalisation in England was much slower and less distinctive.[54]

Notes

1. Here, 'Industrialisation' is seen in the context of economic growth in the modern sense, i.e., a sustained increase in the real per capita product, including structural changes in the economy, increasing investment ratio, technical changes and changes in production functions. See for example, S. Kuznets, *Modern Economic Growth. Rate, Structure and Spread*, London 1966.
2. Statistics on population from B.R. Mitchell, *European Historical Statistics 1750-1970*, London 1975, Table C1.
3. Statistics from G.W. Roderick and M. Stephens, *Education and Industry in the Nineteenth Century*, London 1978, p. 74.
4. All growth figures calculated from Mitchell, op. cit., Table E1. Statistics for England only are not available, while we have to use the UK figures. We assume that these figures reflect the English performance rather well.
5. The word 'Germany' is used throughout as a collective denotation for the German states during the nineteenth century, although it is not formally adequate until 1871.
6. The following examples can be mentioned: L. Playfair, J.S. Russel, Th. H. Huxley, B. Samuelson, M. Arnold, H. Roscoe and Ph. Magnus.
7. See D.S. Landes, *The Unbound Prometheus. Technological Change and Industrial Development in Western Europe from 1750 to the Present*, Cambridge 1972, p. 150.
8. See, for example, R. Torstendahl, *Teknologins nytta. Motiveringar för det svenska tekniska utbildningsväsendets framväxt framförda av riksdagsmän och utbildningsadministratörer 1810-1870*, Uppsala 1975.
9. *Betänkande med utredning och förslag angående den högre tekniska undervisningen. SOU* 1943:34, p. 65.
10. See G. Eriksson, *Kartläggarna. Naturvetenskapernas tillväxt och tillämpningar i det industriella genombrottets Sverige 1870-1914*, Umeå 1978; N. Runeby, *Teknikerna, vetenskapen och kulturen. Ingenjörsundervisning och ingenjörsorganisationer i 1870-talets Sverige*, Uppsala 1976; R. Torstendahl, op.

cit. (1975) and *Dispersion of Engineers in a Transitional Society. Swedish Technicians 1860-1940*, Uppsala 1975.

11. See, for instance, S.F. Cotgrove, *Technical Education and Social Change*, London 1958, p. 202; Ch. Erickson, *British Industrialists. Steel and Hosiery 1850-1950*, Cambridge 1959, p. 35; Ch. P. Kindleberger, *Economic Growth in France and Britain 1851-1950*, Cambridge, Mass., 1964, p. 153; W.J. Reader, *Professional Men. The Rise of the Professional Classes in Nineteenth-Century England*, London 1966, p. 117; P.W. Musgrave, *Technical Change, the Labour Force and Education. A study of the British and German iron and steel industries 1860-1964*, Oxford 1967, pp. 62, 102, 256, 263; and D.C. Coleman, 'Gentlemen and Players', *Economic History Review*, 2nd Series, Vol XXVI:1 (1973), pp. 104, 112.

12. Regarding the initial phase of industrialisation and the shaping of the system of technical education in the individual country, we can, to a certain extent, see a parallel to Gerschenkron's view on European industrialisation during the nineteenth century, on the basis of the country's degree of economic backwardness. See A. Gerschenkron, *Economic Backwardness in Historical Perspective*, New York 1965, p. 353 ff.

13. It should be noted that such corps already existed in France during *l'Ancien Regime*. During the Napoleonic period, however, the French educational system was reformed and given new directions. See p. 30ff.

14. For example, see M. Argles, *South Kensington to Robbins. An Account of English Technical and Scientific Education since 1851*, London 1964. Commenting upon the claim made by the Balfour Committee (Report of Committee on Industry and Trade, HMSO, 1927) that before any progress could be made within the English industry, it had to discover and make known its industrial requirements, Argles states that this apathy partly 'sprang from industry's fundamental disbelief in the value of research and development, which in turn was a consequence of its reluctance to employ scientists and technologists at top level' (p. 70). Concerning management education Argles observes that the formal education of industrial managers in the nineteenth century was almost nonexistent; 'Indeed, in England, this important branch of education did not really get into its stride until after 1945' (p. 125).

15. S. Lilley, 'Technological Progress and the Industrial Revolution 1700-1914', *The Fontana Economic History of Europe*, Volume III, 1973.

16. Ibid., p. 190 and pp. 187, 213.

17. Ibid., p. 194.

18. Ph. Deane, *The First Industrial Revolution*, Cambridge 1969, p. 132.

19. Lilley, S., 1973, p. 212.

20. Ibid., p. 231.

21. Ibid., p. 235.

According to Rolt, up to the middle of the nineteenth century there was a clear distinction in England between 'science' and 'the useful arts'; 'If engineering was a useful art, with the emphasis on art, science was a department of philosophy – "natural philosophy" ' . . . From James Watt onwards, engineers had drawn freely upon scientific knowledge and method – indeed their success would otherwise have been impossible – but they mistrusted scientific theories and formulae, regarding them as no substitute for their practical experience . . . All science was then pure and the phrase "applied science" had no meaning'. L.T.C. Rolt, *Victorian Engineering*, The Penguin Press, Pelican Books 1974, p. 168.

22. However, it should be mentioned that, according to later research, applied science seems to have played a somewhat bigger role in the Industrial Revolution in England than has generally been assumed before. See A.E. Musson (ed.), *Science, Technology and Economic Growth in the Eighteenth Century*, London 1972, p. 56 ff. But the examples which can be cited only indicate slight differences in the

evaluation of the importance of applied science during this phase. The form remains essentially the same and is generally related to the fact that empiricism, trial and error, practical experience, etc. were dominant.

23. See Ch. Babbage, *Reflections on the Decline of Science in England and on Some of its Causes*, London 1830.

24. J. Jewkes, 'How much Science?', *The Economic Journal*, Vol LXX (1960), p. 3. Jewkes discusses the situation in Great Britain during the 1950s concerning the scientists and technologists/engineers.

25. The term *Technische Hochschule* (Germany), *Teknisk Högskola* (Sweden), i.e. technical university, is used throughout for the German and Swedish schools despite the fact that this designation is formally valid only from the latter half of the nineteenth century. For more details, see Chapter 2.1.

26. Nevertheless, certain aggregate calculations are possible on the basis of the French engineers' statistics for the present period (see under 2.2).

27. See K-H. Ludwig, *Technik und Ingenieure im Dritten Reich*, Düsseldorf 1974, p. 19, note 6.

28. The engineers graduated from the Swedish technical universities made up about 40 per cent of the total number of technicians within the private and public Swedish industrial fields. (Calculated from *Utlåtande och förslag till den lägre tekniska undervisningens ordnande*. Committee of 1907. Printed in *Bihang till Riksdagens protokoll vid lagtima Riksdagen i Stockholm 1918*, pp. 485-86.)

29. Ch. P. Kindleberger, 'Technical Education and the French Entrepreneur' in Ed. C. Carter II, R. Foster and J.N. Moody (eds.), *Enterprise and Entrepreneurs in Nineteenth and Twentieth Century France*, Baltimore 1976, p. 13.

30. By 1900 the number of members of the Institution of Civil Engineers exceeded 6 000, and according to the censuses of 1881 and 1911 the number of civil engineers was approximately 7 100 and 7 200 respectively. (Cf. Roderick and Stephens, 1978, p. 130 and W.J. Reader, 1966, p. 211.)

31. At the Berlin *Technische Hochschule*, for example, during the last decade and a half of the nineteenth century an average 9 per cent of the students at the school were of foreign nationalities (calculated from archival material at the school). In this respect the Swiss technical university at Zurich – that was important at a stage in the development of the higher technical education (see p. 33ff.) – had a remarkably high proportion of foreign students and consequently also educated a number of, for example, the highly qualified German engineers. During the 1860s and 1870s, almost 55 per cent of the students at the school were thus foreigners; in the 1880s it was just above 50 per cent, and from the 1890s to World War I the number of foreign students made up 40 per cent of the total. (Calculated from *Eidgenössische Technische Hochschule 1855-1955. Ecole Polytechnique Fédérale*, Zurich 1955, pp. 255-56.) It is noticeable that no Englishmen, at least up to the 1870s, were studying at the prestigious Swiss school, as the proportion of foreign students was very high. The reason was that no students from England had come forward who could qualify for entrance. (See G.C. Allen, *The British Disease*, London 1979, p. 43.)

32. In Sweden at the beginning of the twentieth century, about 10 per cent of the practising technicians and engineers had graduated at foreign technical schools. However, the fact that a certain number of the engineers emigrated seems to counterbalance this to a certain extent. In the case of Sweden, it has been shown that 17.5 per cent of the Chalmers engineers who graduated before World War I had emigrated, but it was assumed that the KTH engineers did not emigrate in such large numbers. Of those who graduated from the KTH in 1888-1895, almost 86 per cent stayed in Sweden or returned home after studies abroad. (See *Betänkande med undersökningar och förslag i anledning av tillströmningen till de intellektuella yrkena. SOU* 1935:52, pp. 252-54.) As Table 2.12 shows,

approximately 7 per cent of the members of the Swedish Association of Engineers and Architects were practising abroad in 1909; of these 7 per cent, about two-thirds remained in Europe.

33. See *International Encyclopedia of Social Sciences*, 1968, Vol. 5, p. 70.

34. F.B. Artz, *The Development of Technical Education in France 1500-1850*, Cambridge, Mass. 1966, p. 161. Regarding England, Artz states that engineering 'was only emerging as a profession' while in other countries it was 'merely a skilled craft'.

35. *International Encyclopedia of Social Sciences*, Vol. 5, p. 70. The word 'engineer' is, however, recorded even in the Norman Chronicles from the twelfth to fourteenth centuries. From the medieval latin term *'ingenium'* meaning machine, ingenious mechanical device, the French term *'engin'* was formed, whence 'engineer' could be derived. See K. Malmsten, 'Från krigsingenjör till bergsingnjör', *Daedalus. Tekniska Museets Årsbok*, Stockholm 1942, p. 60. According to Artz (1966, p. 47) the term 'engineer' was applicable to those who looked to the construction and use of military machines (*engines*). The constructors of French navy vessels were probably the first to be officially awarded the title of engineers; 'Ingenieurs-constructeurs de la marine' in the year 1689. See Malmsten, 1942, p. 65.

36. *Ingeniör Lexikon*, Stockholm 1805, p. 208. In his study *The Mechanicals. Progress of a Profession*, London 1967, Rolt states that the term 'engineer' was known only in its military sense. For example Newcomen (1663-1729) – the maker of steam engines, a mechanical engineer – called himself an 'ironmonger' and Brindley (1716-1772) – the canal builder, a civil engineer – a 'millwright' (p. 3).

37. See Rolt, 1967, p. 6. According to Rolt, John Smeaton was the first person in England to style himself 'civil engineer', in order to distinguish his title from that of the military engineers and not to define his activity.

38. G.S. Emmerson, *Engineering Education: A Social History*, Newton Abbot 1973, p. 7.

39. There are certain similarities between Emmerson's definition and the definition given by W. McClellan in his address to the American Institute of Electrical Engineers, 1913. McClellan argued that engineering came from 'the merger of two distinct traditions, the practical or mechanical on the one hand and the theoretical or scientific on the other'. He also maintained that there were three types of engineers: the applied scientist, the mechanic and the designer. Of these, the designer was 'the real engineer'. W. McClellan, 'A Suggestion for the Engineering Profession' in *The Transaction of the American Institute of Electrical Engineers*, 1913, here quoted from E.T. Layton Jr, 'American Ideologies of Science and Engineering', *Technology and Culture*, Vol. 17:4 (1976), p. 696.

40. *International Encyclopedia of Social Sciences*, Vol. 15, p. 576. The word is derived from the Greek *techne* which means 'art' or 'skill' and *technikos* simply means a person who possesses a certain art. See A. Zvorikine, 'Ideas of Technology. Technology and the Laws of its Development', *Technology and Culture*, Vol. 3:4 (1962), p. 443.

41. In *Victorian Engineering*, Rolt writes in connection with his discussion of the development in England during the second half of the nineteenth century: 'In the process Engineering, which had been hailed as the foremost of the Useful Arts, would come to be referred to as a branch of technology, a horrible but necessary word' (p. 167). Concerning the word technology it has been stated that it 'has come to have so many meanings that it can no longer be precisely defined in a way that conforms to usage. Nor can one objectively circumscribe the term without doing it violence'. H. Popitz, H-P. Bahrdt, *Technik und Sozialarbeit. Soziologische Untersuchung in der Huettenindustrie* (1957), p. 26, here quoted from R. Rürup, 'Historians and Modern Technology', *Technology and Culture*,

Vol. 15:2 (1974), p. 167, footnote 17. But the question is whether it has ever been possible to precisely define the meaning of technology. It is symptomatic that J. Schmookler in his study *Invention and Economic Growth* (Cambridge, Mass. 1966) made 'technology' stand for applied science, engineering knowledge, invention and subinvention (p. 5). See also N. Rosenberg's vague but, according to the author, 'useful and legitimate abstraction': 'Technological knowledge ought to be understood as the sort of information which improves man's capacity to control and to manipulate the natural environment in the fulfilment of human goals, and to make that environment more responsive to human needs.' *Technology and American Economic Growth*, New York 1972, p. 18.

42. See Runeby, 1976, p. 76.

43. B. Abrahamsson, *Militärer, makt och politik*, Stockholm 1972, p. 10. A typological method attempts to find criteria which would enable 'professions' to be distinguished from 'non-professions'.

44. Abrahamsson, 1972, p. 11 ff.

45. Artz, 1966, p. 161.

46. Ibid., p. 251.

47. See K-H. Manegold, *Universität, Technische Hochschule und Industrie. Ein Beitrag zur Emanzipation der Technik im 19. Jahrhundert unter besonderer Berücksichtigung der Bestrebungen Felix Kleins*, Berlin 1970, Chap. II:2.

48. Ibid., p. 59. After Robert von Mohl, *Die Polizeiwissenschaft nach den Grundsätzen des Rechtsstaates*, 2.A., Bd. I, Tübingen 1844, p. 478.

49. See G. Holmberger, *Svenska Teknologföreningen 1861-1911*, Stockholm 1912.

50. See Runeby, 1976, p. 88 ff. This association was of a type similar to the German *Verband deutscher Diplomingenieure*, founded in 1909.

51. After Runeby, 1976, p. 90. Table 1.

52. Holmberger, 1912, p. 120.

53. Runeby, 1976, p. 100.

54. It should only be briefly mentioned here that a process, where technical activities successively developed into a profession, began in England towards the end of the eighteenth century. But as regards the dominant category, mechanical engineering, this process is considered to have taken a long period of time.

2 FRANCE, GERMANY, SWEDEN

2.1 Higher Technical Education – The General Pattern

France pioneered the institutionalised higher education, followed by certain German states, particularly Baden and Prussia. In the development of this kind of education, the following institutions were instrumental in setting a pattern: *École Polytechnique* and its *École d'Application* – particularly *École des Ponts et Chaussées* and *École des Mines* – and *École Centrale des Arts et Manufactures* in France, as well as some German institutions, particularly the technical university of Karlsruhe but also, during the latter half of the nineteenth century, that of Berlin.[1] In this context, we should also mention the Swiss *Eidgenössiches Polytechnicum* in Zurich during a phase extending from the middle of the 1800s (see p. 33ff).

It has been alleged that technical education in these countries was not directly associated with the aims of industrialisation.[2] With regard to eighteenth-century France, one can find support for such a view – of course it boils down to a question of definition – but it is not of the German states, where the express purpose, right from the start, was to improve the nation's manufacturing sector.[3]

In Sweden the German-Austrian system especially influenced the development of education in this field,[4] and the first statutes of the technical university in Stockholm (1826) resembled those of the *Polytechnisches Institut* of Vienna in more ways than one.[5] Among other things, they stated that the purpose of the school was 'to collect and supply knowledge and information necessary in order to run . . . crafts and industry successfully'.[6] It should be noted, however, that there were qualitative differences – to the advantage of Chalmers – in the education provided by the two leading Swedish technical schools up to the middle of the 1840s (see below); this was partly due to different views regarding the content of a technical education on the part of the boards of directors. But it is obvious that the education in both Stockholm and Gothenburg was geared to the society's industrial needs (the implications being less sophisticated at the time, of course); consequently, it affected the process of modern industrialisation in Sweden.[7]

France

Significant advances were made in the French technical education as early as the seventeenth century, but most of them took place during the eighteenth century, and they were of great importance to the present structure of education in France.

This early foundation of technical schools has been seen as a result of mercantilism, the economic policy of the time; 'To make the state strong in a military and naval way, to improve the quality and amount of its manufactures so that they would command both domestic and foreign markets were the steady aims of statesmen. Unless this is clearly grasped, the effects to improve technical education cannot be understood.'[8] However, it should be emphasised that even by the time of the French Revolution, most workers in French industries were still trained by rule-of-thumb methods. It was in the field of the higher technical education that France was in the lead, and that is where important advances were made during the eighteenth century.[9]

The *Academie Royal d'Architecture* (1671) has been characterised as being the first higher technical school in France,[10] while the *École du Genie Militaire* (1745) and the *École des Ponts et Chaussées* (1747, reconstituted in 1775 on a regular basis) have been mentioned as the best technical schools of *l'Ancien Régime*.[11] The latter school has been described as being the first modern civil engineering school in the world.[12] Here we should also mention the *École des Mines* (founded in 1769, it was authorised in 1778 and received its name in 1783), although the education at the time of the revolution was not better than that offered by several mining schools in Germany.[13] Its importance, however, lies in the school's connection with the *École Polytechnique*, founded in 1794,[14] and the established system of specialisation schools, *Écoles d'Application*, which were attended after finishing studies at École Polytechnique. From the point of view of higher instruction in the fields of civil engineering, École des Mines — together with École des Ponts et Chaussées — was thus the most important of these *Écoles d'Application*.[15] During the nineteenth century École des Mines developed into a school of general mechanical engineering. Along with the *École Centrale des Arts et Manufactures*, founded in 1829 on private initiative,[16] it educated engineers to be employed by the government and by private industry.[17]

Although the revolution caused a certain amount of disorder in French education,[18] the long-run effects were favourable ones, as the period stimulated the non-university sector of scientific and technical education. This was a consequence of the philosophy that served as the ideological foundation of the revolution; 'The purpose of the educators, the scientists, and the legislators of the Revolutionary era was to turn the young away from the study of God, of men, and of the past and to direct their attention to nature and science, to what was believed to be socially useful, and to the future.'[19]

Special attention should be paid to the year of the foundation of the *École Centrale des Arts et Manufactures* as the date must be held to be a comparatively late one, considering the early start of the French engineering schools in general. Against the background provided by the purpose of the revolution, it is more noticeable still. But the reason is probably simple: after the outbreak of the revolution, École Polytechnique and its École d'Application and other existing technical schools were supposed to satisfy the demand for engineers.

However, the supply was probably not adequate. This became obvious as the commercial competition, especially with England, increased after the war years; 'By 1820, there were demands for more trained engineers everywhere.'[20] In this context, an argument has been put forward by Comberousse in his history of the École Centrale. He has emphasised the negative effects of the non-existence of a *Corps ingénieurs civils* – i.e. non-military engineers – which still prevailed in France at the time of the Restoration. Such a corps of engineers did, on the other hand, exist in contemporary England, and Comberousse feels that 'C'est là sans nul doute, la vraie cause de la supériorité que l'Angleterre acquit sur les marchés du monde après la paix de 1815.'[21]

At that time, however, the French government took a very small interest in the industrial sector and the leaders of the manufacturing class. Any measures concerning technical education taken by the government involved a concentration on the extant schools. This state of affairs was partly due to the highly conservative, even reactionary, general attitudes held by the restored Bourbon monarchs; another reason may be found in the fact that students of the École Polytechnique and some of the other technical schools were hostile to the government; 'the government seems to have had the notion that all technical education was, in itself, hostile to the rule of "the throne and the altar" '.[22]

Consequently, the initiative of founding a new technical school, the École Centrale, especially intended for the education of non-military engineers, had to be taken by private individuals.[23]

Again it should be emphasised that the engineers examined here are the highly qualified ones, i.e. those intended for leading positions in French industry. We have already mentioned the existence of École

Polytechnique and its Écoles d'Application. However, for those French students who wanted an engineering career in private industry after finishing their studies at secondary schools, there was no institution which could provide the necessary elements of higher instruction; 'Il est donc permis d'avancer que l'enseignement complet des sciences industrielles n'existe pas encore en France . . . c'est l'ambition des fondateurs de l'École Centrale des Arts et Manufactures'.[24]

The aim of the school was 'de former des ingénieurs civils [i.e. non-military engineers], des directeurs d'usines, des chefs de manufactures, des constructeurs, et, en outre, de donner à tous ceux qui veulent prendre part au développement des affaires industrielles l'instruction qui leur manque, soit pour en apprécier la valeur, soit pour en surveiller la marche'.[25]

The existence of an industrial science was clear to the founders and 'de mettre les *pratiques industrielles* à la hauteur des *théories scientifiques*' was the mission of the education offered at the school.[26] Both theory and practice should be taught. Based on the model of the École Polytechnique, the École Centrale was founded as a kind of industrial École Polytechnique.[27]

In the middle of the nineteenth century, Paris was still the leading centre of engineering education, but the long period of French ascendancy approached its end.

While the French engineering schools, especially the École Polytechnique and its Écoles d'Application and the École Centrale des Arts et Manufactures maintained their high intellectual traditions unimpaired, they came to be rivaled by the Swiss Polytechnic School and later by the rapidly advancing schools of Germany and America in enrollment, material equipment, and financial resource. The rising stream of foreign students now went first to Switzerland, then to Germany. The days of outstanding French leadership in all types of technical education were rapidly passing.[28]

Germany

For the German-speaking states, the École Polytechnique was of great importance at first; it also encouraged the establishment of technical schools of a new type, the so-called *Gewerbeinstitut*, later Technische Hochschulen. Such institutions were found in Vienna (1815), Karlsruhe (1825), Berlin (1827) and other cities. The first school of this new type, however, was the one founded in Prague in 1806.[29]

In the pertinent literature, it has been established that in the development of technical education 'zuerst Paris, dann Wien und in weiterem Verlaufe Karlsruhe . . . in der Lehre und Pflege der technischen Wissenschaften, wie auch in ihrer Organisation bahnbrechend einwirkten'.[30]

The *Polytechnische Schule* in Karlsruhe was founded in 1825 as the result of an official planning commission for establishing a polytechnical school. The purpose of the school was expressed in its statutes as follows: 'Sorge für die Bildung unseres lieben und getreuen Bürgerstandes und überhaupt eines jeden, der sich den höheren Gewerben widmen, dazu die nötigen Vorkenntnisse, vorzüglich aus der Mathematik und aus den Naturwissenschaften sich erwerben und deren unmittelbare, in das Einzelne gehende Anwendung auf die bürgerlichen Beschäftigungen des Lebens kennenlernen will.'[31]

Thus it is evident that the purpose was to provide a 'necessary' education for occupations in manufacturing and industry, and that this education should basically comprise mathematics and the natural sciences.

The Karlsruhe Polytechnic soon — in the early 1830s, to be exact — began to serve as a model of other German polytechnics. In 1832 the school was reorganised and the most important aspect of this reorganisation, led by the Cabinet minister Nebenius, was the introduction of the principle of the *Fachschule*, an element that has been held to herald a new era in the history of higher technical education.[32]

From the 1840s onwards, the scientific approach — in general and basic subjects as well as in applied ones — that had characterised the Karlsruhe Polytechnic from the outset grew even more accentuated. The development of the basic principles of machine building founded on mathematics turned the subject into a completely new science. Thus, on the initiative of F. Redtenbacher[33] (1809-63) in 1847, the *höhere Gewerbeschule*, one of five *Fach* schools, was divided into one mechanical-technical and one chemical-technical school. In 1860, the first of these was given the name *Maschinenbauschule*.

During his last six years, Redtenbacher was the headmaster of Karlsruhe Polytechnic. It was during the time of his successor F. Grashof (1806-93) that the idea of a technical university, which emerged during Nebenius' time and developed further during that of Redtenbacher, was brought to its completion.[34] As the initiator of the *Verein Deutscher Ingenieure*, and the editor of its journal, Grashof must be considered a very important person in the development of the interests of engineers and their profession.

In an 1864 speech, he stressed that the polytechnical schools should strive to develop a *Hochschule* character and that the education needed for simple or moderately advanced technical jobs should be supplied by public schools. Grashof summarised the aim and character of the poly-technical schools as follows:

> Sie sei eine Technische-Hochschule und bezwecke die den höchst-berechtigten Anforderungen entsprechende wissenshaftliche Ausbild-ung für diejenigen technischen Berufsfächer des Staatsdienstes und der Privatpraxis, welche die Mathematik, die Naturwissenschaften und die zeichnenden Künste zur Grundlege haben sowie auch die Ausbildung von Lehren der an der Schule vertreten technischen und Hilfswissenschaften.[35]

With regard to this development, Karlsruhe was also in the lead,[36] and from the reorganisation of 1865 the school received the character of Hochschule. Its status was then on a par with that of a university; it was hence largely autonomous and had its own administration. How-ever, it did not receive the name of Technische Hochschule (TH) until 1885, and the right to confer doctor's degrees (*Promotionsrecht*) was not granted until 1900.

The pattern of higher technical instruction established in Karlsruhe came to be followed – after certain changes inspired by *Eidgenössische Technische Hochschule* in Zurich, founded in 1855[37] – at all the German technical universities. Therefore, the main importance of the Berlin *Technische Hochschule* – so named after 1879, but in practice a Hochschule since 1866[38] – lies in the quantitative aspect; for instance, from 1871/72 until the turn of the century nearly 30 per cent of all the students at German technical universities studied at Berlin TH.[39] But Berlin TH was important from certain qualitative points of view as well. Firstly, the Hochschule ideas developed in Zurich probably to a certain extent found their way to Germany via Berlin. Secondly, the Berlin TH enjoyed considerable prestige in the united German state and was thus in a position to recruit leading teachers and professors. In this context it should also be stressed that during the latter decades of the nineteenth century, Berlin TH played the leading role in the struggle of the technical universities to obtain equal status with the universi-ties.[40] Emmerson in his *Engineering Education* characterises the opening of the Polytechnic School of Berlin at Charlottenburg as 'the crowning glory' of a 'careful and purposeful education in science and technology';[41] and in *The Organisation of Science in England*, Cardwell emphasises that in England 'the great polytechnic at Charlottenburg' — besides the industrial and university chemical laboratories — was 'one of the most admired of German institutions at the time',[42] that is, around the turn of the century.

Franz Reuleaux, one of the best-known and most influential German engineers, would appear to have constituted a personal link of some importance, connecting Karlsruhe, Zurich, and Berlin. A student of Redtenbacher's, he was called to a Chair in *Mechanik und Maschinenkonstruktion* in Zurich. After a successful period of work in Switzerland, Reuleaux was director of the Berlin *Gewerbeakademie* for eleven years, from 1868 onwards. After the amalgamation of the *Bauakademie* and the Berlin Technische Hochschule, he became the headmaster of the institution.[43]

Important as such personal links were, the same principles were established at all the German technical universities, and it seems more reasonable to emphasise the importance of the Berlin technical university on the basis of its size, localisation, and prestige.[44]

As was mentioned above, Berlin spearheaded the demand for equal status between the Technische Hochschulen and universities towards the end of the nineteenth century; this was a long-drawn struggle which cannot be considered as resolved until the turn of the century, when the right to confer doctor's degrees was obtained.

Sweden

Formally, there was only one technical university in Sweden during the nineteenth century, namely *Kungl. Tekniska Högskolan* in Stockholm, which received that name in 1877 after fifty years of existence. In actual fact, however, it is relevant to speak of two,[45] since Chalmers in Gothenburg provided an education of almost equal standing, qualitatively speaking, to the Stockholm school. According to the organisation plan of 1876, the upper division at Chalmers should take a middle-rank position between the technical university of Stockholm and the technical secondary school, founded around 1850.[46]

During the first half of the nineteenth century, as was pointed out before, the education at Chalmers was clearly of a more scientific type than the training offered in Stockholm. Still, the Chalmers school — its upper divisions, that is — did not obtain the name *Chalmers Tekniska Högskola* until 1937. Two years later, joint statutes for the technical universities in Gothenburg and Stockholm were proclaimed.

The technical university in Stockholm, which was given its first statutes in 1826 under the name of the Technological Institute, grew out of the Mechanical School, founded in Stockholm in 1798. It goes back to the *Laboratorium Mechanicum* (later the Royal Model Chamber) founded by Christopher Polhem – 'the father of technical education' in Sweden[47] – in 1697.

However, the education provided at the Institute was, as was pointed out above, extremely elementary during the first two decades. In fact, the original statutes stipulated that it should 'generally' be popular and practical rather than severely scientific.[48]

Quarrels within the Institute concerning its main function, the sort of education it was to offer, and its training methods soon arose. The headmaster of the school, G.M. Schwarz – 'Sweden's first "teknolog" ' – was against anything to do with theory, arguing that the education should be absolutely free from anything that could be called 'scientific'. Despite work by committees and proposals for improvements, the Institute had to wait until the middle of the 1840s for a radical change in a specifically technical scientific direction, with mathematics as the basic subject. New statutes, passed in 1846, took effect in 1848; and it is from this time onwards that the school may be considered as an institution for higher technical education.[49] According to its statutes, the Institute should be 'An institution of education for those young men who pursue some kind of industrial occupation which cannot be properly performed without knowledge of nature, chemical and mechanical technical knowledge especially.'[50] From that time onwards mathematics was the basic subject at the Institute. In the school's further development during the century, the move of the upper division of the Falu Mining School to Stockholm, and its amalgamation with the Technological Institute, was of significant importance. In 1867, it resulted in the new statutes, in which the scientific character of the education was explicitly emphasised; 'for young men who want to pursue some kind of technical profession, requiring a sufficient *scientific education*'.[51]

However, the struggle to obtain the same status as the classical university was a protracted one in Sweden, too. Not until 1927 did the technical university in Stockholm receive the right to confer doctor's degrees – the symbol of equal standing – while Chalmers had to wait another decade.

2.2 Number of Engineers

The problem of analysing the development of highly qualified engineers in France involves the total view of the annual number of graduates (flow) and thereby the problem of calculating the total number (stock) of these engineers in the society at different times. As regards the French engineers' choice of occupation after completing their education, information is available for certain schools. With regard to Germany and Sweden, such information is only available to a limited extent. The statistics concerning the supply of highly qualified engineers in the latter countries are better, which makes it easier to calculate the total number.

France

As has been indicated above, the French engineering schools have a long tradition. Of the 145 schools operating in the middle of the 1970s, about 10 per cent were founded during the first half of the nineteenth century or earlier, 20 per cent having been established during the second half of that century. Hence, almost one third of the schools that educate the *Ingénieurs diplomés* were founded before the twentieth century.[52] The entrance qualifications for most of the schools are very high – with École Polytechnique, École des Mines (Paris) and École Centrale on top – and have been so for a long time.

An estimation based on the present (1974) annual number of graduates[53] from schools existing at certain earlier dates – which entails an exaggeration, of course – indicated that around 1850 the annual supply was about 2 200 and around 1900, 4 700 engineers. These figures overestimate the true situation for 1900 by about 50 per cent, and the figure for 1850 should be reduced by about 75 per cent.[54] Thus, the flow of diploma engineers in mid-nineteenth-century France was about 550 and around 1900, 2 000-2 500. However, engineers with lower technical qualifications – for example those educated at the *École des Arts et Métiers* – are also included in these figures, and the annual number of graduates with a higher technical education at the middle of the nineteenth century was only 250-300; around the turn of the century, there were probably only about 1 000.[55]

If we assume that, on the high side, the total number of these engineers, i.e. the stock, was 1 500 around 1800, the number around 1850 would be about 6 500.[56] In the mid-60s, the total figure was about 10 000, and by the end of the 1880s it was twice that amount. Around the year 1900 there were almost 30 000 highly qualified engineers in the French society, and at the outbreak of World War I approximately 40 000 engineers had this educational background (see Table 2.1 and Figure 2.1).

As a proportion of the economically active population in the 1850s, engineers constituted about 0.8 per thousand. In the middle of the 1890s the figure was 2.0 per thousand, and in the beginning of the 1910s about 3.0 per thousand.[57]

Table 2.1: Number of Engineers with Higher Technical Education. France, Germany and Sweden 1850/54 — 1910/14; Calculated Annual Averages

	France	Germany	Sweden
1850/54	7 100	4 000	700
1855/59	8 200	5 600	800
1860/64	9 500	7 900	900
1865/69	11 100	10 500	1 000
1870/74	12 700	13 200	1 200
1875/79	14 700	19 600	1 300
1880/84	17 000	26 600	1 400
1885/89	19 700	30 300	1 500
1890/94	22 800	34 000	1 700
1895/99	26 500	41 500	2 000
1900/04	30 800	47 300	2 400
1905/09	35 200	54 200	2 900
1910/14	40 600	62 600	3 300

Comment: For the method of calculation, see notes 56 (France), 59 (Germany) and 62, 63 (Sweden).
Annual figures are shown in the Appendix.

Germany

The number of German students in higher technical education shows a marked and fairly steady increase during the nineteenth century. In the beginning of the 1830s the number was about 500 and around 1850 about 1 000. It reached 5 000 in the beginning of the 1870s, 10 000 in the middle of the 1890s, and the turn of the century, 1900, it amounted to 15 000. The number levelled out after that; it is actually possible to note a decrease. In 1914 the number of matriculated students at the *Technische Hochschulen* was about 11 500.[58]

Information about the number of students every year at these schools enables us to calculate the annual number of graduates approximately, and the stock of these engineers, at various times.[59]

Figure 2.1: Number of Engineers with Higher Technical Education.
France, Germany and Sweden 1850/54 — 1910/14

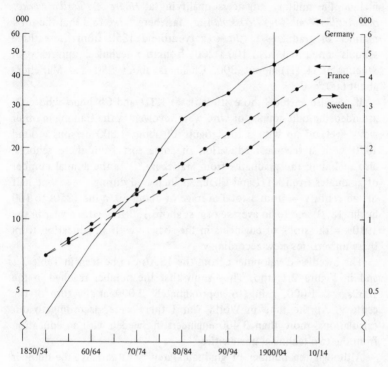

The result of these calculations can be seen in Table 2.1 and Figure 2.1.

In the middle of the nineteenth century the total number was only about 3 500; twenty years later, though, united Germany had about 12 000 highly qualified engineers. During the latter part of the 1880s there were, as the Table indicates, more than 30 000 practising engineers with this education in Germany, and at the time of the War the figure was more than 60 000.

Relatively speaking, 2.0 per thousand of the economically active male population were engineers with these qualifications in the early 1880s.[60] In the mid-1890s it was 2.4 per thousand, and some years before World War I it was of a similar magnitude — about 3.0 per thousand — to that in France.

Sweden

For obvious reasons, the total number of students in Sweden qualifying at the corresponding Swedish technical schools in Stockholm and Gothenburg was smaller than in France and Germany. If we also include the military engineers qualifying at *Högre Artilleriläroverket och Artilleri- och Ing. Högskolan* at Marieberg,[61] we find that the total number of graduates in the society around 1850 from these three schools amounted to 1 100-1 200; from the technical university in Stockholm (KTH) about 300, Chalmers about 250 and Marieberg about 600.[62]

If we also include those students at KTH and Chalmers who only attended the odd course, or who were involved in the training in other ways, we end up with a total figure of about 1 800 persons around 1850 with a technical education of some sort from these schools, also including the graduates from Marieberg.[63] As the annual number of graduates from KTH and Chalmers increased during the second half of the century – from a total average of about 30 in the 1850s to 100 in the 1890s and to an average of just above 150 graduates a year in the 1910s – the stock of engineers in the society with an education from these universities grew accordingly.

The Swedish development from the 1850s can be seen in Table 2.1 and in Figure 2.1, too. They show that the number reached in the 1860s was 1 000, rising to approximately 2 000 at the turn of the century. At the time of World War I there were, according to our calculations, more than 3 400 engineers in Sweden with an education from the two technical universities.[64]

Although the number of graduates grew substantially, the number of qualified engineers in the society as a proportion of the economically active male population was, however, lower than in France and Germany. Not until the time of World War I did Sweden possess the same relative number of qualified engineers – 2.0 per thousand – that Germany had as far back as the early 1880s and France in the mid-1890s. In relative terms, the Swedish development thus deviated somewhat from the Continental one.

Our comparison of the French and German situation, involving the aspects of annual supply (flow) as well as total number (stock), showed significant differences, especially in absolute terms. While the total number of highly qualified engineers in France around 1850 was, according to our calculations, almost twice the number in Germany, we find that from the early 1870s the number is larger in Germany than in France and that the differences during the nineteenth century kept growing in Germany's favour. Around the turn of the century as well as at the time of World War I, the number of highly qualified engineers in France was only two-thirds of the German figure.

2.3 Demand for and Careers of the Engineers

As was indicated in Chapter 1, industry's demand for qualified engineers and the engineer's choice of career are essential components in an analysis based on the figures presented above. The aggregate figures for France conceal the crucial fact that compared to Germany and Sweden, only a small number of engineers were employed within industry, particularly in private industry.[65]

France

As was pointed out above, there was a lack of qualified engineers within the private industry in France during the first half of the nineteenth century. The purpose of École Centrale des Arts et Manufactures was to rectify this situation. With regard to the private industry, École Centrale also came to provide the dominant proportion of those individuals in management and supervisory functions who had a qualified engineer's education.

From the opening of the École Centrale, its graduates were also to be found in all sectors of French industry and commerce.[66] As they — in accordance with the purpose of the school — to a large extent held leading positions in the firms where they were active, it seems plausible that they also played an important role in developing the French private industrial and commercial sectors. Artz has stressed that 'The École Centrale supplied the higher theoretical knowledge without which manufacturers and their chief aides could not direct their establishments or control the managers and foremen whom they employed.'[67]

Table 2.2 shows the occupations in which the graduates of the École Centrale 1829-51 were found.

Table 2.2: Occupation of Graduates from the École Centrale, 1829-1851

	Number	%
Agriculture	18	1.7
Architecture	39	3.7
Railroads	119	11.3
Teaching	42	4.0
Textile manufactures	36	3.4
Public works	53	5.0
Industrial chemistry	57	5.4
General civil engineering	56	5.3
Machine manufacture	30	2.9
Metallurgy and mining	79	7.5
Commerce and miscellaneous manufactures	522	49.7
	1 051	

Source: Artz, 1966, pp. 251-2.

Besides the obvious fact of the dispersion of engineers to all economic sectors, it should be noted from the table that the biggest single one was the railway sector. The importance of the graduates of the École Centrale to the development of the French railways has received due attention in relevant literature. It was found that 163 former students of the school had been working in this field during the years 1835-63, and of these 28 were directors and chief engineers, 79 *ingénieurs principaux*, and 56 ordinary engineers.[68]

On the whole, between 1829 and 1885 no less than 27 per cent went to the railways and 9 per cent to the public works. Forty per cent were active within industry, and 16 per cent as consulting engineers. Four per cent were in the teaching profession.[69]

The importance of the École Centrale engineers in leading positions in, for example, the manufacturing and mining industries, becomes even more obvious from Table 2.3, which shows the situation in 1863.

Table 2.3: Practising Engineers in 1863. Graduated at the École Centrale

	Number	%
Head of forge or owner of mine	124	24.9
Owner of big enterprise/firm	68	13.7
Production/design of machines	54	10.8
Spinning mill	43	8.6
Producer of chemical products	38	7.6
Agriculture	37	7.4
Entrepreneurs; public works	35	7.0
Director and owner of gas-work	31	6.2
Production of sugar	28	5.6
Director of glass-work	23	4.6
Producer of paper	17	3.4
	498	

Source: Comberousse, 1879, p. 251.

Comberousse, in his history of the school, also claims that 'Ils [the graduates of the École Centrale] formaient comme l'élite de l'armée du travail, et contribuèrent largement à éviter au pays une crise redoubtable.'[70] The potential crisis he refers to was connected with the Anglo French Commercial Treaty of 1860, which was strongly opposed by French industrialists.[71] Still, thanks to the contribution of the engineers of the École Centrale, it was, according to the author, possible to transform French industry rapidly. It has also been emphasised that 'There is reason to believe that the total effect of the liberalising of trade and this piece of governmental assistance to industry [loans from the state to facilitate modernisation and re-equipment] was to permit allround increase in efficiency and some lowering of production costs.'[72]

Of course, it is impossible to establish the importance of leading engineers in this process with any degree of precision. However, it is known that, in French industry during the 1860s, 'On the whole the stronger firms were strengthened, the less competitive went to the wall.'[73]

Table 2.3 indicates that a rather large proportion of the École Centrale engineers was probably found in successful firms. At the Paris Exhibition of 1867 (see also p. 79ff) France and the École Centrale met with considerable success.[74]

L. Playfair, who acted as a judge at the exhibition, was asked by Lord Taunton, Chairman of the English Schools Inquiry Commission, to note down the substance of a conversation between them about technical education. Among other things, Playfair referred to Dumas – 'well known as a *savant*, and who, from his position as a Senator of France, and President of the Municipal Council, has many opportunities of forming a correct judgement' – and wrote: 'Dumas . . . assured me that technical education had given a great impulse to the industry of France. In going through the Exhibition, whenever anything excellent in French manufacture strikes his attention, his invariable question is, "Was the manager of this establishment a pupil of the École Centrale des Arts et Manufactures?" and in the great majority of cases he received a reply in the affirmative.'[75]

Thus the information studied supports the view that the graduates of the École Centrale were, from an early stage, important to the development of the French economy. Levy-Leboyer, among others, has pointed out that École Centrale facilitated the diffusion of advanced technology within the industry and contributed, especially after 1860, to the introduction of new talents to the business world. Former students of the school controlled 70 per cent of *la Société des Ingénieurs civils de France*, founded in 1848, opening an employment office for the students of the school in 1862.[76]

Writing in 1879, Comberousse could establish that since the start of the school more than 7 000 persons had studied at the École Centrale, and of these more than 4 000 had finished their studies with a diploma or certificate.[77] Large as this number seems, the figure 7 000 is, for instance, only one-third of the number of students at Karlsruhe Technische Hochschule during the same period.

One objection which could be raised here is that it was not only École Centrale that educated qualified engineers for the French industry. École Polytechnique and its Écoles d'Application educated a large number of engineers, too. However, a study of the choice of careers of Polytechnique and Mines engineers does not alter the impression of a small number of qualified engineers within the French mechanical industry, particularly within the private industry.

The proportions of former students of École Polytechnique within private and public sectors during the first half of the nineteenth century are to some extent discussed by Artz. The information concerning the practising engineers – about 3 500 – within the public sector 1795-1835,[78] shows that about 50 per cent of the Polytechnique graduates entered the artillery and slightly more than one quarter military engineering. Less than 10 per cent entered the corps of mines, bridges and highways.

The proportions of different kinds of engineering do not reflect the attractiveness of different activities, though. In his book about French technical education, Artz refers to an American author (Barnard), who wrote about the École Polytechnique in the early Second Empire. According to this author, the students selected state services in the following order: first, roads and bridges and mines (very nearly on an equality); second, munition manufactures; third, naval architects; fourth, army engineers; fifth, artillery; sixth, general staff; seventh, hydrological corps; eighth, tobacco administration; ninth, telegraph; tenth, general navy; and eleventh and last, naval artillery.[79]

Although a large part of the graduates chose positions in the public sector, they were not compelled to do so. But usually pressure was brought to bear on the graduates with a view to channelling them into government services. 'That more students did not choose positions in private industry seems to have been due to the extraordinary *esprit de corps* the school had built up.'[80]

It can be established that up to the middle of the 1830s two thirds of the *polytechniciens* were in governmental services. Accordingly, one third entered the private sector,[81] i.e. were so-called *démissionaires*.

Consequently, the private sector's demand for qualified engineers seems not to have been satisfied. Despite the fact that about half of the students in, for example, École des Mines and École des Ponts et Chaussées were not graduates of the École Polytechnique, and that these so-called *externes* usually found employment within private industry, 'the supply of trained engineers after 1815 always remained insufficient'.[82] As was emphasised above, the founding of the École Centrale des Arts et Manufactures, by a group of private capitalists, should be seen in this context.

The Polytechnique graduates' choice of career 1870-1914 is shown in Table 2.4.

Table 2.4: Polytechnique Graduates' Choice of Career. Total, within Corps Militaire and within Corps Civil de l'État, 1870-1914

Corps Civil	Total		Corps Militaire			Corps Civil		
	No.	%		No.	%		No.	%
(governmental)	1 132	12.2	Engineers[a]	419	5.7	Mines	190	16.8
Corps Militaire	7 323	78.8	Officers[b]	6 904	94.3	Ponts &		
						Chaussées	749	66.2
Demission	840	9.0				Divers[c]	193	17.1
TOTAL	9 295		TOTAL	7 323		TOTAL	1 132	

Notes:

[a]Génie Maritime, Artillerie Navale, Poudres et Salpêtre, Hydrographie.
[b]Artillerie, Génie, Marine, Infanterie Coloniale, Aéronautique, Commissariat de Marine.
[c]Eau & Forêts, Corps des Manufactures d'État.
Source: C. Mercie, *Les Polytechniciens 1870-1930, Recrutement et Activités,* 1972 (unpublished). Extracts from Tables 3, 4, 5, pp. 20, 22, 23.

It is evident that a very small proportion entered the private sector directly after finishing studies at the École Polytechnique; during these years not one out of ten graduates chose this career.

Within the public sector, military service attracted a majority – 87 per cent – of the graduates, and the majority of the polytechniciens in the military sector – almost 95 per cent – joined the armed forces and became officers.

Civil engineering was the dominating profession within the public non-military sector. In these years, almost two-thirds thus continued their studies at École des Ponts et Chaussées. Only a minor proportion of those in *Corps civil* seems to have been attracted by production and design, i.e. mechanical engineering.

The graduates of the École des Mines, like those of the École Polytechnique, entered both the public and private sector. This was also considered as the school's dual aim: 'préparer, d'une part, les ingénieurs recrutés à l'École Polytechnique qui doivent former le Corps national des mines, et, d'autre part, instruire les jeunes gens qui veulent obtenir le diplome d'"ingénieur civil des mines" que confère l'École'.[83]

Around 1830 the proportion of the graduates entering the private sector was 10 per cent, in 1840 20 per cent and in the middle of the 1850s 30 per cent. The proportion fluctuated between these figures until·1918.[84]

It has been emphasised that graduates of the advanced École des Mines at Paris were much sought after and either became government mining inspectors or worked as engineers for private industry.[85] The position of those Mines engineers in the early 1880s and in the beginning of the twentieth century who were admitted to the school as students a mere ten years before is shown in Table 2.5.

As was the case with the graduates of the École Centrale, the graduates of the École Polytechnique and the École des Mines were to be found in all industrial and occupational sectors of the French society, where they rapidly attained leading positions. However, as was indicated above, only a small proportion chose the private sector, and to the very great proportion in the public sector, a military career was more attractive than one in engineering. For those who chose an engineering career, civil engineering seems to have been more attractive than mechanical engineering.

The total number of graduates from the École des Mines, and the number of demissionaires from the École Polytechnique, were probably too small an addition to the former students of the École Centrale to refute the impression that the number of qualified engineers in French mechanical industry during the nineteenth century was indeed small.

Table 2.5: Engineers' Socio-Professional Function, Ten Years after Admission to the École des Mines, 1870s, 1890s. Per cent

Year of Admittance	Cadres Moyens	Cadres Supérieurs	Cadres Dirigeants	Patrons
1871, 1872	88	4	8	–
1891, 1892	72	9	15	3

Cadres Moyens: Engineers.
Cadres Supérieurs: Engineer 'capable of organising the fabrication and the work and of assuming the responsibilities connected with direction'. (After G. Depeux, *La Société Française, 1789-1960*, 1960, p. 251): *L'ingénieur divisionnaire – l'ingénieur principal, chef-adjoint, le chef de section, l'ingénieur en chef*, etc.
Cadre Dirigeant: takes the large technical, commercial, financial or social decisions.
Patrons: Entrepreneurs, industrialists, owners of firms.

Source: E. Baucher, A. Moore, *La Formation et le Recrutement de l'Ingénieur Civil des Mines, 1817-1939*, 1973 (unpublished). Extract from Table at p. 142.

In this context, we should also note that Levy-Leboyer, after deducting those students who chose a military career, found that the annual average number of engineers graduating from École Polytechnique, Mines, Centrale and École du Genie Maritime during the 1850s, 1870s and 1890s was only about 125, 210 and 275 respectively. For the years 1900-14, the average had gone up to 325.[86] A calculation based on the figures provided by Levy-Leboyer showed that the number of engineers from these schools being added to the industry during 1850-1914 was only about 8 700.[87]

Germany

In Germany, the situation was significantly different. This proposition can be based on our knowledge of the large number of students at the technical universities.[88] However, as Lundgreen has emphasised in his study *Techniker in Preussen während der frühen Industrialiseriung* (1975) − the period from the 1820s to the 1870s − technical education does not by itself create a modern industry. There must be a demand for technicians, too.[89] As for Prussia, it is obvious from the Lundgreen study that there existed a large, although differentiated, and growing market for these.[90] Given the structure of the Prussian industrial economy, he establishes that 'Das Angebot an staatlich ausgebildeten Technikern mochte . . . in Quantität und Qualität mehr oder weniger optimal sein.'[91]

There is no reason to doubt that the situation in other German states was different; at least that there was a market for these engineers.

There exists some qualitative evidence regarding the demand for qualified engineers at the middle of the nineteenth century. For example, Playfair in his *Industrial Instruction on the Continent* (1852; mentioned above) pointed out the large demand for these engineers. Discussing the Berlin *Gewerbeinstitut*, he said that 'the students are in great demand by manufacturers, and it is rare to find men who go out with good class certificates waiting any considerable time for employment'[92] and about the polytechnic school of Karlsruhe, Baden: 'The formal certificates of the Special Technical schools are said to be in the highest estimation, and command immediate employment to the possessors.'[93]

In his discussion concerning the German development from the 1870s to the inter-war period of the twentieth century, Landes could establish: 'Most important, entrepreneurs prized the graduates of these institutions [the Hochschulen and universities] and often offered them respected and often powerful positions − not only the corporate giants with their laboratory staffs of up to a hundred and more, but the small firms also, who saw in the special skills of the trained technician the best defence against the competition of large-scale production.'[94]

Pfetsch, among other authors, has emphasised the considerable and increasing importance of science and technology during the nineteenth century as an explanation of the German economic growth.[95] In this context, he has also stressed the close connection between the development of the German technical instruction system and the governmental science policy, as well as indicating how the latter's role changed during the century.[96] He sees governmental activities in the technological sphere at the beginning of the nineteenth century as a sequel of a mercantilistic policy and holds that these were not initiated by demands coming from economic interests. From the middle of the century, the state became more and more 'neutral', and towards the end of the 1800s it merely responded to the demands of industry; the initiative lay with the industrial interests.[97] From then onwards, according to Pfetsch, government activities in the field of technical advance played a catalytical role in the growth process.[98] He thus shows that public investments (1882-1914) in research and technology ran in a countercyclical manner, while the number of matriculated students at the Technische Hochschulen (1869-1914) responded to fluctuations in the business cycle.[99]

Supposing that Pfetsch's proposition about the governmental response to the demands of industry is correct, and considering the sensitivity of TH students to the fluctuations in the business cycle, one might argue that there was probably a balance on the market for qualified engineers from the latter part of the nineteenth century.[100]

Sweden

As was noted in the previous chapter, the density of highly qualified engineers in Sweden was lower than on the Continent. However, the reason for this was not attributable to a lack of demand for engineers — and as was pointed out by the Committee of 1890 for the expansion and reorganisation of the technical university in Stockholm (KTH) — the result of a continuous stream of inventions and new fields of application for the technical sciences had created a 'continuous increased need for persons with a thorough technical education.'[101] That the demand for engineers was very high during the latter decades of the nineteenth and early-twentieth centuries has been stressed, too.[102] According to an investigation in 1919 in order to find out the distribution of engineers in various branches and the need for engineers in

the Swedish economy, the existence of an 'unmet need' was established; so was, generally, an increasing shortage of qualified engineers.
The small increase in the number of employed engineers with an
education from KTH and Chalmers was particularly noticeable with
regard to the period studied (1910-19).[103]
Concerning the careers of the Swedish engineers during the first
half of the nineteenth century, information is scanty. It is only for the
first decade of Chalmers engineers (1829-39) that their occupational
fields after finished studies are known. The engineers' dispersion to all
types of occupational fields is clear from Table 2.6, but it also shows
that the dominating area in which the engineers were practising are
industry and the crafts. Taken altogether, 56 per cent of the Chalmers
engineers were active in these fields.

Table 2.6: Occupation of Chalmers Engineers. Graduates 1829-1839

		Number	%
Industrial fields:			
Manufacturing and maintenance of machinery and instruments; metal processing	41		
Milling, iron and mining work	16	79	38.9
Chemical industry, pharmacy	15		
Manufacturing business (unspecified)	7		
Trade		45	22.2
Crafts		34	16.7
Navigation		18	8.9
Military		17	8.4
Continued studies		6	3.0
Agriculture		4	2.0
		203	

Source: Bodman, 1929, p. 190.

Where later periods are concerned, better information is available. As
has been pointed at before, the occupational fields of those engineers
qualified at the KTH 1850-80 have been studied, as well as the technicians employed in the Swedish economy in 1908.[104] Table 2.7 shows
the result of these studies.
It is obvious from the Table that the qualified engineers were active
within all fields of the economy. Stress should be laid on the fact that a
large proportion were active in mining, and in mechanical and other
industrial fields: approximately 45 per cent of the KTH engineers in

1880, and in 1908 more than 60 per cent. Almost three-quarters of the Chalmers engineers were in these fields at the latter date. To these percentage figures, we should also add the rather larger proportion of the engineers in the railway sector, who were thus active in building up the country's infrastructure.

Table 2.7: Occupation of Graduates from the KTH and Chalmers in 1880 and 1908

Industry/Field of Occupation	KTH 1880	%	1908	%	Chalmers 1908	%
Mining	62	8.4	132	13.6	29	5.7
Engineering	172	23.2	254	26.1	199	39.4
Wood, Pulp and Paper	24	3.2	57	5.9	46	9.1
Chemical	23	3.1	42	4.3	15	3.0
Mineral	2	0.0	34	3.5	34	6.7
Other industries[a]	42	5.7	87	9.0	45	8.9
Building and Construction	56	7.6	105	10.8	81	16.0
Railways	119	16.1	196	20.2	48	9.5
Telegraph	6	0.0	32	3.3	5	1.0
Various[b]	234	31.6	33	3.4	3	0.6
	740		972		505	

Notes:

[a] Including food, textile, leather and rubber, consultation and unspecified.
[b] Including national defence, state administration, education, others and unknown occupation.

Sources: *Industritidningen Norden* 1881, p. 39ff, C.G. Rystedt's article 'Teknologernas verksamhet och öden' (Rystedt's register of KTH graduates 1850-1880); 'Den lägre tekniska undervisningen i Sverige' (The lower technical education in Sweden) *Bihang till Riksdagens Protokoll* 1918, p. 485 ff.

The 1908 study also shows that the qualified engineers to a large extent held leading positions in the fields in which they were active. This fact is most pronounced for the KTH engineers. A classification in three different levels was made in the investigation concerning job positions: Level I, the highest, included management, Level II was a middle-range position, and Level III included foremen and similar employees. The result is seen here in Table 2.8.

Table 2.8: Job Position in 1908. Engineers Graduated from the KTH
and Chalmers

	KTH		Chalmers	
	Number	%	Number	%
Level I	352	36.2	142	28.1
Level II	451	55.7	263	52.1
Level III	79	8.1	100	19.8
	972		505	

Note: Level I denotes independent leading positions in management and similar,
Level II positions at a dependent and middle range, where a qualified technical
education is required. Level III includes persons in lower positions, such as foremen, draughtsmen, etc.

Source: See Table 2.7. 'Den lägre tekniska undervisningen i Sverige', p. 518. For
the criteria of classification, see pp. 487-8.

The studies of Swedish engineers have not considered the age structure of the individuals, i.e. the pattern of careers over time. However,
this aspect has been thoroughly studied by Torstendahl in his *Dispersion of Engineers in a Transitional Society. Swedish Technicians 1860-
1940* (1975). The career of the technician was studied at three definite
points in time: 1-2 years after graduation, then 14-16 years after and
finally some 29-31 years after passing the finals. The main investigation
comprised the graduates from the schools 1881-1910 – a total of
2 987 highly qualified engineers; 2 015 from KTH and 972 from
Chalmers.[105] Certain results from the Torstendahl study that are of
importance in the present context can be seen in Table 2.9.

Table 2.9: Engineers Graduated from the KTH and Chalmers 1881-1910,
Employed in Metal Industry, Building Trade, Consultation Engineering
and Railway Service at Different Stages in their Careers. Per cent

	Percentage of graduates in these branches after certain years 1-2 (C1), 14-16 (C2) and 29-31 (C3)		
	C1	C2	C3
KTH	71.1	61.5	58.7
Chalmers	68.2	63.2	59.2

Source: R. Torstendahl, *Disperson of Engineers*, 1975, p. 99.

The dominance of those four most frequent branches that were presented in the Table is obvious, as well as the fact that these branches were dominant over time, too, i.e. the engineers to a very large extent stayed in the industrial branches in which they started their career. The study also indicates that these engineers rather quickly reached leading job positions.

Table 2.10 shows the individual careers in the private sector.

Table 2.10: Positions in Private Business, after a Certain Number of Years in a Career. Engineers graduated from the KTH and Chalmers 1881-1910

Percentage of Group at Work in Private Business

	Business Leaders	Employees	Consultant Engineers	Unspecifiable	Group Size
1. 1-2 years					
KTH	2.8	80.6	16.3	0.3	870
Chalmers	1.0	88.0	10.6	0.5	407
2. 14-16 years					
KTH	27.8	54.4	16.8	1.0	915
Chalmers	19.8	63.9	15.4	0.9	460
3. 29-31 years					
KTH	40.5	40.5	18.3	0.6	820
Chalmers	32.7	47.3	18.0	2.0	395

Source: Torstendahl, *Dispersion of Engineers*, 1975, p. 217.

That a dominating proportion of the highly qualified engineers were active in the private sector is also obvious from the Torstendahl study and here in Table 2.11.

Table 2.11: Distribution of Engineers into Public and Private Sectors after Certain Years from Graduation. Graduates of the KTH and Chalmers 1881-1910

	Public Sector	Private Sector	Non-Sectorised Specified Occupation	Unspecified Occ.	Total Number
1. 1-2 years					
KTH	32.7	62.0	1.0	4.2	1 378
Chalmers	23.6	72.7	1.3	2.5	556
2. 14-16 years					
KTH	37.3	57.2	3.1	2.4	1 558
Chalmers	24.7	69.6	2.1	3.5	662
3. 29-31 years					
KTH	38.0	54.3	5.5	2.2	1 476
Chalmers	25.8	64.2	6.5	3.5	598

Source: Torstendahl, *Disperson of Engineers*, 1975, p. 155.

However, it should be noted that the minor public sector recruited a slightly increasing proportion of the engineers as their careers proceeded. It is a fact that the engineers graduated at Chalmers were more frequently active in the private sector than those graduated at the Stockholm technical university. As can be seen in Table 2.12, the percentage figure for the Chalmers engineers is similar to the picture of the occupational sectors supplied by the members of *Svenska Teknologföreningen*, STF around 1910 — approximately 70 per cent in the private sector.

Table 2.12: The Swedish Association of Engineers and Architects. Distribution of Members in Fields of Occupation According to the 1909 Register of Members

Occupation or Activity	In Sweden	Abroad		Total
		In Europe	Non-Europe	
In governmental service[a]	318	—	—	318
In municipal service[b]	150	—	2	152
Ironworks and mining	154	7	4	165
Other industries	507	45	27	579
Engineers in private business[c]	151	4	5	160
Architects in private business[d]	93	—	—	93
Private railways and trams	47	—	—	47
Teachers	69	—	—	69
Insurance	25	—	—	25
Businessmen	13	—	—	13
No information[e]	174	24	7	205
	1 701	80	45	1 826

Notes:

[a] Public railways, telegraph, marine corps, civil engineers, the board of pilotage, patent and registration office and hydro-electric power board.

[b] Engineers for public works or geodetic assessment and town architects.

[c] Keeper of patent agency and entrepreneurs in civil engineering.

[d] All architects in non-public sector.

[e] Including retired engineers and members on travels for the purpose of studying.

Source: Holmberger, 1912, Bil. VIII, p. 266.

The general conclusion concerning the careers of the highly qualified Swedish engineers from the latter half of the nineteenth century —

besides the large demand for them — is the engineers' dispersion to all fields of occupation, but first of all to the industrial sector, a predominant proportion going into the private sector. In accordance with the contemporary idea of a hierarchical system in industry and in the economy at large, these engineers also to a very large extent — and fairly rapidly — reached leading positions.

A comparison between France, Germany and Sweden from the point of view not only of the total supply of highly qualified engineers, but also of their choice of careers, indicates that in France only a comparatively small number of them was to be found in industry during the nineteenth century.

A discussion of the social standing of the engineers, especially in France, and the attractiveness of an engineering career in industry is important in this context (see Chapter 1, p. 18) and will be considered in the following section. The German and Swedish cases will be dealt with first.

2.4 Social Standing of the Engineer

According to the economic-historical literature, people in managerial and entrepreneurial functions in Germany during the nineteenth century had their social origin in the middle class,[106] and Kocka in his recent 'Entrepreneurs and Managers in German Industrialization' — which is based on a thorough investigation of the literature on the subject — has stated how 'the overwhelming majority' of the entrepreneurs of the industrial revolution in Germany came from 'industrial and business trades and activities'.[107]

It has been shown that students at the schools that educated people for the above-mentioned functions belonged to the same group; concerning Baden during its early industrialisation by Fischer,[108] and regarding Prussia by Lundgreen.[109] Commenting on the statistics on the Polytechnical school at Freiburg in 1818/19, Fischer writes, 'Interessant ist die soziale Herkunft der Schüler. Die Väter gehören zum grössten Teil *den* mittleren Bürgerschichten an, für die die Schule ausbilden soll',[110] while Lundgreen summarises the information about the social background of students at the Berlin Gewerbeinstitut from the early 1820s, to the early 1870s as follows: 'ein relativ hoher Anteil von Handwerkersöhnen, vielleicht über 50%, in den ersten drei Jahrzehnten, bei insgesamt kleinen Studentenzahlen und einer kleinen Minderheit von Nicht-Stipendiaten; nach 1850/55 eine etwa proportionale

Verteilung der Studenten auf Söhne von Beamten, Kaufleuten/Fabrikanten und Handwerkern, bei steigenden Studentenzahlen und einer kleinen Minderheit von Stipendiaten . . .'.[111] He also emphasises that a noticeable – but, according to him, probably not surprisingly – large proportion of sons of the 'höheren Stände' could be seen after the 1850s.[112] Certain results from the Lundgreen study are shown in Table 2.13.

Table 2.13: Social Background of Students at the Berlin Gewerbeinstitut 1855-1867

Father's Occupation	1855 %	1861 %	1867 %	Total
Craftsman, Employee (lower)	21.9	21.1	31.7	25.9
Merchant, Manufacturer, Employee (higher)	23.4	28.5	28.3	27.4
Civil servants, Officers	36.3	34.8	24.4	30.5
Liberal professions, Others	18.4	15.6	15.6	16.2
Number	201	365	442	1 008

Source: Lundgreen, 1975, p. 120.

As far as we know, no such studies exist regarding the subsequent development during the nineteenth century.

However, since the social status of higher technical education grew, an increasing number of people with this kind of training is likely to have come from the so-called upper class. This would, in that case have been due to what has been called 'die Invasion der materiellen Interessen' or the 'realistischen Tendenz'[113] in the German society of the nineteenth century, and Landes, among others, has pointed out how 'A veritable cult of *Wissenschaft* and *Technik* developed' in Germany.[114] Thus, an important factor in the social, as well as political, standing and strengthening of the engineering profession in Germany during the nineteenth century – which most German authors seem to agree happened during the latter part of the century[115] – was the recognition of the importance of *Technik* as a cultural factor, existing alongside, and in conjunction with the humanistic disciplines and the pure sciences.

As Manegold emphasised, the development of the technical universities was, in this context, inevitably bound up with the question concerning the social status of the *akademische* technicians. This question
was the corollary of another problem, that of the status of the technical
universities compared to the universities; 'Gleichstellung der Technischen
Hochschulen mit den Universitäten, Gleichachtung der technischen
Wissenschaften, Gleichberechtigung der realistischen Bildung, das alles
sollte für die Ingenieure vor allem auch eine entsprechende dienstliche
und gesellschaftliche Ebenbürtigkeit in der traditionellen akademischen
Welt bedeuten, ihre Integration in den Kreis der älteren "gelehrten"
Berufe und ebenso eine entsprechende Stellung im staatlichen Leben,
in Regierung und Verwaltung.'[116]

Since the industrial strength of Germany was indisputable, and the
highly qualified engineers could be characterised as 'Pionier deutscher
Geltung und Kultur',[117] the technical universities' claim for equal
status with the universities was intensified.

At the centenary celebration of the Technische Hochschule in Berlin
in 1899, Wilhelm II could say: 'Das Ansehen der deutschen Technik ist
schon jetzt sehr gross. Die besten Familien . . . wenden ihre Söhne der
Technik zu, und Ich hoffe, dass das zunehmen wird.'[118] Of course, an
expression like 'the best families' is very vague, but some kind of social
criteria must have been involved; consequently, something was said
concerning the standing of technical education.

As has been pointed out above, the right to confer doctor's degrees
which was granted to the technical universities at the turn of the
century entailed the formal and official recognition of the equal status
of these schools as compared with the universities.

Concerning the social background and status of the engineers in
Germany, it can thus be said that they mainly came from the middle
class, but that a large proportion belonged to the upper class as early as
the middle of the nineteenth century. In this context, understanding
Technik as a cultural factor was important; this in turn was closely
linked with the German industrialisation. The engineer was considered
as the synthesis of *Teknik und Kultur*; 'die Idee der Maschine und die
Idee des Menschen mussten zu der höheren Einheit des Ingenieurs
zusammentreten!'[119] The highly qualified engineer was at the head of
this process. Thus, the equal status of the technical universities with
the classical university was of great social importance, not only for the
highly qualified engineer but also for engineers with lower education.
The whole question must be seen in the context of the ascending
bourgeoisie during the nineteenth century, especially during the second

half of the century. It reflected the spirit of enterprise in society, a new *Weltanschauung*, and a new political vision.[120]

The Swedish development was similar to the German one in many respects. We have seen that the ideas concerning technical education that stemmed from Germany were more or less imitated in Sweden, at least from the middle of the nineteenth century, and also that the process of professionalisation of the highly-educated engineers was similar to the German development.[121]

Parallel to the Swedish industrial development, the prestige and social standing of the engineers was growing. With his technical qualifications, the engineer came to represent a new elite and in a way became the symbol of the industrial epoch and a new culture; the engineer was, to quote Eriksson, 'standing at the line where industrialization and science met'.[122]

As was the case in Germany, the striving for the right to confer the doctor's degree to those with a higher technical education was the culminating point in the ascending social process. By such means, the qualified engineer was to receive his deserved social respectability, and also protection against the less qualified. Runeby has convincingly interpreted these views as a misunderstandable expression of the very close relationship between scientific striving and academic ideas of status and professionalism; 'The men of the "new" elite sought their social acceptance by joining the "old" one.'[123]

One way of establishing the interest in – and, by the same token, also to a certain extent the social respectability of – a qualified engineering education, is to take a closer look at the competition for places at the relevant schools.

Regarding the period 1850/54–1910/14, we find that an average of some 70 per cent of the applicants to the KTH were admitted, but also that the situation before the 1880s is very difficult to assess.[124] However, from that time onwards – in the early 1880s nearly 85 per cent of the applicants were admitted – a falling trend in the percentage of admitted students is noticeable, and in the period 1910/14 less than 60 per cent of the applicants to the KTH were admitted.

However, although these figures can in part be seen as a result of a growing social standing of the profession, it should be emphasised that the attractiveness of a qualified engineering education was to a large extent determined by the country's economic situation, i.e. depended on growth and fluctuations in the business cycle. This has also been pointed out.[125]

The background of economic conditions must also be taken into account when looking at the intended careers of students after finishing secondary school (*studentexamen*), although it seems reasonable to argue that the social factor is more important in this situation. Table 2.14 shows the situation towards the end of the nineteenth century and during the first decade of the twentieth century.

Table 2.14: Swedish Students — Intended Careers after Secondary School ('Studentexamen')

Year	Technical Education	%	University	%	Total
Average per year					
1884-89	39	5.8	324	48.0	672
1890-99	84	13.3	248	40.3	620
1900-09	135	13.6	365	35.9	1 005

Source: *BiSOS* Education, Public secondary schools for men.

The Table thus indicates a growing proportion of students who aimed towards a career as a *civilingenjör*, while at the same time a pronounced falling trend of those who intended to begin their studies at the classical university is noticeable.

The figures also seem relevant for the true situation, since it can be shown that during the period 1900-14, an average 11.5 per cent of the total number of students who left the secondary schools with a Certificate — and who were consequently in a position to choose between the classical university and, for example, a technical university — were admitted to the technical university in Stockholm or Gothenburg.[126] Knowing as we do that about 40 per cent of the students at public secondary schools belonged to the highest social group around the turn of the century, and more than 50 per cent to the middle class,[127] it might well be argued that a higher technical education was socially respectable.

Statistics concerning the socio-economic background of the students at the KTH and Chalmers are not easily available for the nineteenth century and as far as we know no study considering that aspect for the century exists.

However, there is no reason to assume that the pattern of the socio-economic background in Sweden deviated from the German one, i.e. that the students at the technical universities were largely recruited from the middle class, but also to a growing extent from the highest social group.

A study concerning matters such as the social recruitment of students to universities and various specialised schools after *studentexamen* in the first quarter of the twentieth century confirms this assumption.[128] The result of the study with regard to the technical and classical universities at the turn of the century and in the mid-1920s is presented in Table 2.15.

Table 2.15: Students Who Started their Studies at the Swedish Technical and Classical Universities after Finishing Secondary School, 1902/04 and 1924/26. Students Divided According to Father's Occupation

Social group	KTH/Chalmers		Lund/Uppsala Univ.	
	1902/04 %	1924/26 %	1902/04 %	1924/26 %
(a) I	50.2	60.2	55.6	43.9
(a) II	43.4	26.9	37.9	44.1
III	6.4	12.9	6.5	12.0
(b) I	39.4	40.1	51.9	38.7
(b) II	54.1	47.0	41.6	49.3
III	6.4	12.9	6.5	12.0
Number of Students:	279	394	1 121	2 025

By calculations according to the (a) method engineers and accountants belong to social group I, while according to (b) they belong to social group II.

Source: P. Dahn, *Studier rörande den studerande ungdomens geografiska och sociala härkomst*, Lund 1936, Table 148, p. 385.

The Table shows the very similar pattern of students' social background at the classical and technical universities. At the turn of the century, social groups I and II made up almost 95 per cent of the total number, and irrespectively of methods of calculation 40-50 per cent of the students at the KTH and Chalmers belonged to the upper class — a proportion which was 40-60 per cent by the mid-1920s.[129] Only about 5 per cent came from the lowest social ranks at the turn of the century. In the 1920s this proportion had doubled, both at the technical and the classical universities.

Students with a higher technical education in France were drawn from the middle and upper social classes. On the whole, the proportion of students from the latter was, for the nineteenth century, greater than in Germany and Sweden. This is indicated by Table 2.16.

Table 2.16: Social Origin of Students at the École Polytechnique, École des Mines and École Centrale. Per cent

	École Polytechnique 1800-1870	1870 1900	École des Mines 1890-1914	École Centrale 1830-1900	
Rentiers, Proprietors	30.1	19.1	9.8	31.8	
Entrepreneurs, Wholesalers[a]	13.8	16.2	14.3	34.6	
Liberal professions	32.0	26.4	41.7	12.7	
Milieux aisés	*75.9*	*61.7*	*65.8*		*79.1*
Employees[b]	18.1	20.6	22.7	10.9	
Craftsmen, Shopkeepers[c]	4.2	12.0	9.3	5.4	
Peasants, etc.	1.8	5.7	2.2	4.6	
Milieux modestes	*24.1*	*38.3*	*34.2*		*20.9*
Total	100.0	100.0	100.0	100.0	

Notes:
[a] Indust. négociants.
[b] Employés subalt.
[c] Artisans, commerc.

Source: Extract from Levy-Leboyer, 1974, p. 25.

It can also be seen that 60-80 per cent of the students at the Polytechnique, Mines and Centrale during the century belonged to social classes which were considered well off. What does this fact — in combination with previously established evidence about graduates' choice of careers — indicate concerning the social standing of the engineering profession in France?

It was mentioned before that the entrance qualifications to the French engineering schools are very high. In the 1970s, for example, only 5 per cent of the applicants to the École Centrale and less than 2 per cent of would-be École des Mines (Paris) students were admitted.[130]

The qualifications required during the nineteenth century were also high, if not of the same order. At the École Polytechnique the number of applicants and admitted students between 1794 and 1836 was 14 782 and 5 628 respectively, and in the latter year the number of candidates was six times the number of places open.[131] Total figures of the number of applicants admitted between 1800-49 and 1850-94 respectively show that during the first period less than one out of three was admitted, and during the second it was only one out of five.[132]

At the École Centrale in 1859 one of two candidates was admitted[133] and during the 1860s and 1870s two out of three. In 1900, only one in three applicants[134] for a place at the Centrale could start their studies.

As a general characteristic of the diploma engineers, it has been said that they 'en France gardent un prestige certain, d'ailleurs lié au prestige spécifique de certaines Écoles et à la qualité du recrutement qui en résulte'.[135] It is said that there also exists a prestigious ranking order of the *Grandes Écoles*: 'Une classification des écoles d'après leur "image de marque" et leur prestige existe, sans aucun doute, chez tous les français.'[136]

This transpires from the students' choice of schools at the *concours*.[137] Laffitte, in his study of the French engineering schools in the beginning of the 1970s, considered that in a comparison of the different *concours*, 'on constatera par exemple qu'il existe des démissions entre Centrale, les Mines et les Ponts avec une tendance à préférer les Mines, et une égalité Ponts-Centrale'.[138] He also argues that 'L'échelle de prestige des écoles pour les employeurs, comporte par la force des choses une corrélation avec l'échelle des prestiges pour les étudiants.'[139]

The high standing of these schools — now as well as in earlier days — seems indisputable. Although it was pointed out above that the period of French ascendancy in engineering education at the middle of the nineteenth century was drawing to a close, this did not mean that the intellectual traditions at the schools were impaired. The social standing of an education at the École Polytechnique, at its Écoles d'Application, or at the École Centrale des Arts et Manufactures, was also unimpaired during the nineteenth century. The designation 'ancien élève de l'École Polytechnique', for example, had constituted — and still did in the beginning of the twentieth century — a title of honour[140] and Kindleberger has stressed that the prestige of the school was 'enormous' and to some extent became 'an end in itself'.[141] But what about the standing of the engineering profession as such in the French society?

Landes has emphasised that the businessman has always held an inferior place in the French social structure.[142] He has pointed out three major forces that were conducive to this situation. Firstly, from the start — which Landes probably sets somewhere during *l'Ancien Régime* — the businessman was detested by the nobility, who saw in him a subversive element. The revolutions of 1789 and 1830 reinforced this attitude. It is also said that 'the entrepreneur was considerably influenced by the prestige of this "superior" group'. That this was indeed the case is obvious 'from his continued efforts to rise into its ranks, either directly or through marriage'.[143]

Secondly, the non-business elements of the bourgeoisie accepted the opinion of the aristocracy. Within all the professional groups that constituted the bourgeoisie, the businessman was generally placed at the bottom of the ladder. Behind this classification one could also discern the tension between the old and new bourgeoisie, a tension that was strengthened by the Revolution; 'the older bourgeoisie, dominated by civil servants and the liberal professions, tended to stress their prestige in the face of capitalist elements'.[144] These aspects of status were also strengthened by considerations regarding the security of professional positions, as well as by the French educational system. 'For these reasons the best talents in France almost invariably turned to the traditional honorific careers such as law, medicine, or government.'[145] Thirdly, Landes also holds that the pressure of literary and artistic opinions significantly supported 'the forces of aristocratic snobbery and bourgeois aspiration'.[146]

Although Landes does not speak about the engineer explicitly, it is among the categories of businessmen and entrepreneurs — Landes alternates between these expressions — that the engineers will be found.[147] Thus, accepting Landes' proposition about and explanations concerning the French social structure, it is not difficult to understand, for example, the polytechniciens' choice of careers during the nineteenth century. On the subject of the French bourgeoisie and its mistrust of innovation, Kindleberger made the following remark in his work on French and British economic growth during 1851-1950: 'Typically the graduates of the École Polytechnique or the École des Mines go into the government or the army. Those who for any reason enter business are suspiciously regarded.'[148]

However, although this sentence probably summarises society's view of the engineering profession very well, it needs to be qualified when considered in connection with the French industrial performance. First, the great importance of governmental activity in the growth process, especially in building up the country's infrastructure, should not be overlooked.

In the debate on French economic growth and the role of the government, two views have been distinguished: one which argues that government promoted industrialisation, which was responsible for economic growth; another that French governments redistributed rather than created. Still, Kindleberger feels that the views can perhaps be reconciled historically.[149] Thus, especially under the Second Empire, but also under the July Monarchy, 'the government played an important role in building the social overhead capital on which economic growth – not just industrialization rested'. On the other hand, during a period from the mid-1880s[150] to the Monnet plan of 1945, the role of government was limited to redistribution.[151]

A large proportion of those polytechniciens who joined the governmental Corps civil were active in building up the French infrastructure during the nineteenth century. A contemporary author cited above writing during the early Second Empire, said that the students' first choice of state service was roads and bridges, besides mines. During the period 1870-1914, no less than two-thirds of those 12 per cent that entered the Corps civil were found in civil engineering (*Ponts et Chaussées*; see Table 2.4). The favourable implications are obvious.

A second qualification refers to what has been called the Cartesian tradition in French inventions and innovations, with its theoretical and scientific elegance.[152] This tradition of technical virtuosity 'is embodied in the heritage of Saint-Simon, which calls for planning, technical decision, order imposed from above, and rapid change'.[153] During the Second Empire these ideas were very evident in the policy of Napoleon III, but not subsequently – not until after World War II.[154]

Landes mentions the fact that France has always produced pioneers, but he also adds that France has had an equally extraordinary talent for putting such men in their place.[155] Gerschenkron – who does not see adverse social attitudes toward entrepreneurs and entrepreneurship as a major retarding force in the development of European countries during the nineteenth century[156] – has commented on this in the following terms: 'In order to maintain his thesis, Landes has to relegate vast and most significant fields of French entrepreneurial endeavor, such as railroads, mines, the iron and steel industry, automobile production, banks and department stores, to qualifying footnotes and dependent clauses.'[157]

However, the statistical information about practising Polytechnique, Mines and Centrale graduates also supports the proposition about the pioneering spirit in French industry. It was emphasised above that the graduates of these schools have always been found in all sectors of French industry and commerce, as well as the fact that they were much sought after and rapidly reached leading positions once they had embarked on a career. Their importance to the development of French industry was also emphasised at an early point. The French 'entrepreneurial endeavor' in several fields might hence partly be explained by these facts – assuming that the polytechniciens, the graduates of École des Mines and École Centrale represented the tradition of technical virtuosity. The fact that the so-called neo-Saint-Simonians during the post-war period were often men of good technical education, many of them graduates of the École Polytechnique, has been mentioned in the literature,[158] and it is well known that many of the most distinguished French innovators were graduates of the Grandes Écoles.

Consequently, it must be stressed that these graduates were important in building up the infrastructure. Furthermore it may well be argued that they were to a large extent responsible for the pioneering tradition in French industry. Still, to mention an example, the polytechniciens' choice of careers – few in production and design, especially in private industry – indicates that the prestige of the engineering profession as such, and in general, was not particularly great; it was the standing of schools like École Polytechnique, École des Mines or École Centrale des Arts et Manufactures that was high. We find indications to this effect in the above-mentioned *esprit de corps* of the graduates from various Grandes Écoles, as well as in the fact that established firms were, generation after generation, identified with, for example, École Polytechnique and École Centrale.[159] The point is – as Kindleberger has expressed it – that in France scientific and technical education were approved for the elite.[160]

Notes

1. It should be mentioned that the *Polytechnical Institutes* of Prague and Vienna were important in the development of higher technical instruction in the beginning of the nineteenth century. See also p. 33, note 30.

Concerning Berlin it should be noted that, according to an investigation from 1847, Berlin did not have a higher technical school. See F. Schröder, *Die höheren technische Schulen nach ihrer Idee und Bedeutung*, Braunschweig 1847, here from G. Grüner, *Die Entwicklung der höheren technischen Fachschulen im deutschen Sprachgebiet*, Braunschweig 1967, p. 17.

However, a slow development towards *Hochschule* forms had begun after 1845 (Grüner, 1967, p. 30) and basic changes were effected during 1849-50, but until then the instruction in Berlin was considered to be on a lower level; 'Ganz und gar ungleich dem Unterricht in Karlsruhe, Prag und Wien, deren Aufbau und Methode stark vom Geist der Pariser Theoretiker durchsetzt war, blieb in Berlin der handwerksmässige Charakter bestehen.' *Die deutschen technischen Hochschulen. Ihre Gründung und geschichtliche Entwicklung,* München 1941, p. 29.

2. See R. Torstendahl, *Teknologins nytta*, p. 13.

The author also takes into account the development in the Austrian double monarchy, Scandinavia and England. Unfortunately, Torstendahl does not attempt to define industrialisation, but in the introductory chapter he alludes to 'the transition from pre-industrial to industrial society' (p. 9). This would indicate an interpretation in line with our view. (See Chapter 1, note 1).

3. For example, see the purpose of the Karlsruhe *Polytechnische Schule* as it was expressed in its statutes (p. 33).

Landes, among others, has stressed how 'the Germans developed their schools [technical as well as general] in advance of and in preparation for industrialization' (1972, p. 348).

It has also been said that the establishment of the German *Gewerbeinstitut* owed something to the spirit of competition among the German states. See K.W. Hardach, 'Some Remarks on German Historiography and its Understanding of the Industrial Revolution in Germany,' *The Journal of European Economic History*, Vol. I (1972), p. 77.

4. See P. Henriques, *Skildringar ur Kungl. Tekniska Högskolans Historia*, Stockholm 1917, p. 55; A. Wijkander, *Chalmerska Institutet 1829-1904*, Göteborg 1907, p. 5; G. Bodman (ed.), *Chalmers Tekniska Institut. Minnesskrift, 1829-1929*, Gothenburg 1929, p. 4.

In his plan (1851) for the Swedish technical education, Wallmark wrote: 'In Germany the industrial education in general is most complete and it is from there we can take the best pattern for the technical institutions.' L.J. Wallmark, *Om tekniska elementar-skolors inrättande i Sverige*, Stockholm 1851, p. 7. See also Runeby, 1976, p. 118 and Torstendahl, *Teknologins nytta*.

5. See Henriques, 1917, p. 101.

6. See Henriques, 1917, p. 95.

7. See Chapter 1, note 2. In his work, Torstendahl has applied two perspectives to his analysis of the purpose of technical knowledge and technological education: 1) technical knowledge within the framework of the existing society, and 2) in the perspective of industrialisation from a sociological point of view, involving indirect effects on society at large as a result of changes in products and production processes.

In the case of Chalmers the latter perspective could, according to Torstendahl, be noted at an early date, but not in case of KTH – until the Technical Committee published its report on the school in 1844. Then, Torstendahl states, 'Technical education should contribute to industrialization – that was the theme.' (*Teknologins nytta*, p. 211.)

However, although it was never explicitly stated, the Committee on the Stockholm institute of 1833, for example, started off with the idea that the education was to satisfy industrial needs – which Torstendahl establishes (p. 210) – but in order to reach this goal, the committee wanted the training to include more theoretical instruction than the trial-and-error type of education favoured by the head of the school. On this basis Torstendahl draws his conclusion regarding the relevant perspective for Stockholm.

But the two perspectives seem to constitute a rather dubious and artificial distinction from an economic-historical point of view, as the content of the technical education is the only criterion on which to found an assessment of the relevant perspective. As was emphasised above, the purpose of both schools, in Stockholm as well as in Gothenburg, was to improve Swedish craft and industry, a process integral to industrialisation in its early stages.

8. Artz, 1966, p. 56.

9. Ibid., p. 110.

10. Ibid., p. 33. The school was as much a school of engineering as of the fine arts.

11. Ibid., p. 151.

12. Ibid., p. 110.

13. Ibid., p. 86.

14. The school was founded as the *École des Travaux Publics*. It received its present name in 1795.

15. Artz, 1966, p. 244.

16. The school was taken over by the government, as from 1857.

17. The bureau of tests for mineral substance founded at *École des Mines* in 1845 should be mentioned here, too; it was a forerunner of industrial research laboratories in engineering schools. See Artz, 1966, p. 85 n. 47, following W.E. Wickenden, *A Comparative Study of Engineering Education in the United States and in Europe*, Lancaster. Pa., 1929, p. 11.

18. For example, *École des Mines* was closed 1790-94 and again 1802-16. See Artz, 1966, pp. 85, 86.

19. Artz, 1966, p. 180. The fact that science was considered as the cornerstone of the new revolutionary education has been stressed by among others, M. Bradley. See her 'Scientific Education for a New Society. The École Polytechnique 1795-1830', *History of Education*, 1976, Vol. 5:1 (1976), p. 11.

20. Artz, 1966, p. 248.

21. Ch. de Comberousse, *Histoire de l'École Centrale des Arts et Manufactures*, Paris 1879, p. 6. See also pp. 24-25.

22. Artz, 1966, p. 248.

23. Four distinguished names should be quoted in this context: the chemist Lavallée, the permanent secretary of *l'Academie des Sciences* Dumas, Péclet, innovator of the *'physique industrielle'* and Olivier, geometrical scientist and a student of Monge. See *L'École Centrale, Annuaire 1976*, pp. 36-37. Lavallée, 'the principal capitalist involved' (Artz, 1966, p. 249), became the first director of the school, the other three being its first teachers.

24. Comberousse, 1879, I, p. 27. The author does not mention the small number of *élèves externes* who were taken on at the *École des Mines*.

25. Ibid., p. 28. See also pp. 110, 322.

26. Ibid., p. 108. *Pratiques industrielles* and *theories scientifiques* are italicised in the source.

27. Ibid., pp. 16 and XIII. The following words were chosen as a motto: 'La Théorie doit éclairer la Pratique avec autant de soin que la Pratique doit vérifier la Théorie.' The second director of *École Centrale* (Perdonnet) talked about the school as *la Sorbonne industrielle*.

28. Artz, 1966, pp. 267-8.

Compare Fohlen's view – a dubious one, in my opinion – that the French technical schools 'served as a model in other countries until the beginning of the twentieth century, when they were eclipsed by the German Technical High Schools'. C. Fohlen, 'The Industrial Revolution in France 1700-1914', *The Fontana Economic History of Europe*, Vol. IV:1, p. 24.

29. W. Treue, 'Das Verhältnis der Universitäten und Technischen Hochschulen zueinander und ihre Bedeutung für die Wirtschaft', in F. Lütge (ed.), *Die wirtschaftliche Situation in Deutschland um die Wende vom 18. zum 19. Jahrhundert*, Stuttgart 1964, p. 224.

30. E. Zöller, *Die Universitäten und Technische Hochschulen*, Berlin 1891, p.59. See also Manegold, 1970, p. 44. Karlsruhe's role as 'Beispiel der Geschichte der heutigen Technische Universität' as well as being 'der ältesten deutschen höheren technischen Unterrichtsanstalt im modernen Sinne' has also been emphasised by, among others, F.R. Pfetsch, *Zur Entwicklung der Wissenschaftspolitik in Deutschland 1750-1914*, Berlin 1974, p. 135.

In *Skildringar ur Kungl. Tekniska Högskolans Historia* (Stockholm) by P. Henriques, published in 1917, it is stated that after *Ecole Polytechnique* certain higher institutions of technical education constituted milestones and models for the modern technical university, namely 1) *Polytechnisches Institut* in Vienna 2) *Polytechnische Schule* in Karlsruhe, 3) *Eidgenössiches Polytechnicum* in Zurich, 4) The American technical universities (for the practical-scientific education) and 5) the then, [i.e. around 1917] rather similarly organised German technical universities (1917, p. 16).

While *École Polytechnique* has been characterised as a top faculty for mathematical and natural sciences, at the *Polytechnische Institut* in Vienna – from 1872 *Kaiserl. Königl. Hochschule* – particular attention was paid to the chemical and mechanical technology. Both basic and applied subjects were taught. (See Henriques, 1917, p. 17 ff.)

It should be pointed out, however, that the reason why the technical schools in the German states usually followed the pattern of the *Wiener Polytechnikum* and not the French pattern has not been much debated in the literature. Maybe the reason is self-evident and Treue has given the correct answer: 'In den deutschen Ländern bestand kein Bedürfnis nach einer grossen zentralen Anstalt, wie die École sie bildete, sondern in diesen industriell relativ rückständigen Bundesländern viel stärker ein solche nach mehr oder weniger gehobenen Fachschulen . . .' (Treue, 1964, p. 225). It has also been asserted that Germany had no important scientific tradition to fall back upon and at the beginning of the nineteenth century was behind the western European countries in scientific knowledge. See Manegold, 1970, p. 18, after F. Schnabel, *Deutsche Geschichte im 19. Jahrhundert*, Freiburg 1954, Bd III, p. 167.

31. F. Schnabel, 'Die Anfänge des Hochschulewesens', in *Festschrift anlässlich des 100jährigen Bestehens der Technischen Hochschule Fridericiana zu Karlsruhe*, Karlsruhe 1925, p. 27.

32. Henriques, 1917 p. 24. See also note 30 in the present study. Besides the preparing school and the mathematical classes, tne Karlsruhe polytechnic consisted of five such *Fach* schools: *Ingenieurschule, Bauschule, Forstschule, Höhere Gewerbeschule* and *Handelsschule*.

33. See Schnabel, 1925, p. 37.

34. Ibid., p. 40.

35. Ibid., p. 41.

36. See Manegold, 1970, p. 73 and Treue, 1964, p. 232.

37. Its character of a technical university was heavily emphasised along with the strict realisation of the principle of the *Fachschule*. However, in order to counterbalance the danger of supplying too narrow an education in the specialisation chosen, lectures and courses dealing with the humanities and with political science were supplied alongside the instruction in the 'pure' sciences.

It should be pointed out that the technical university in Zurich also had Karlsruhe as a model at the time of its establishment. See Manegold, 1970, p. 55 and also pp. 56 and 58.

38. The history of the Berlin *Technische Hochschule* goes back to the *Berliner Bauakademie*, founded in 1799. The purpose of this institution was to supply the theoretical and practical education needed to produce good land surveyors, civil engineers, etc. in governmental service. From the beginning the *Bauakademie* had an exclusive character – it was a school for *Baubeamte*; 'ihr Lehrbetrieb gleich dem einer Universität'.

The Berlin *Technisches Institut* was opened in 1821 and was called *Gewerbeinstitut* from 1827 onwards. Now training in mechanical and chemical engineering was introduced. The establishment of the school was the outcome of the work done by, among others, the *Technische Deputation für Gewerbe* 'zur Beförderung des Gewerbefleisses in Preussen'.

In 1866 the *Gewerbeinstitut* was named *Gewerbeakademie*, and ten years later an amalgamation of the *Bau-* and *Gewerbeakademie* was decided. (Historical notes from Grüner, 1967, and *Die Technische Hochschule zu Berlin 1799-1924. Festschrift*, Berlin 1925.)

39. Calculations are based on the statistical information in Manegold, 1970, p. 321. Manegold's source is W. Lexis, *Das Unterrichtswesen im Deutschen Reich*, Berlin 1904, Vol. IV.

40. See Manegold, 1970, pp. 67 and 82 and Treue, 1964, p. 232.
41. Emmerson, 1973, p. 89. In this context he also mentions the University of Strasbourg in the 1880s. Both are characterised as 'unquestionably the most elaborately equipped colleges of science and technology the world had seen'.
42. D.S.L. Cardwell, *The Organisation of Science in England. A retrospect*, London 1957, p. 150. Cardwell also points out that according to *Reports of the Royal Commissioners on University Education in London* 1910-1913, it was decided that the Imperial College 'must be the London Charlottenburg'.
43. Manegold, 1970, pp. 57-8.
44. With regard to Berlin, Pfetsch has pointed out: 'Wie Paris in Frankreich, so wurde in allgemeinen Berlin als das erstrebenswerte Ziel der wissenschaftlicher Karriere angesehen.' 1974, p. 237.
45. The present discussion does not include the *Högre Artilleriläroverket* at Marieberg opened in 1818 – from 1866 *Krigshögskolan* and in 1878 *Artilleri- och ingenjörshögskolan* (but see p. 40) – although the school has been considered the only real technical university in Sweden up to the reorganisation of the KTH in 1876 (see T. Gårdlund, *Industrialismens samhälle*, Stockholm 1942, p. 228) and a Swedish counterpart to the *École Polytechnique* (see Wallmark, 1851, p. 6). The reason for not taking the school into account here is its marked military orientation; it aimed to be a higher educational institution for officers of all branches of the army. (See P. Sylvan and O. Kuylenstierna (eds.), *Minnesskrift med anledning av K. högre artilleriläroverkets och krigshögskolans å Marieberg samt Artilleri- och ingenjörshögskolans etthundraåriga tillvaro, 1818-1918*, Stockholm 1918, p. 17.)
   However, it should be noted that during the years 1842-1869 the school also educated 'civil' engineers, i.e. non-military engineers, for 'public works' (ibid., p. 20), a course of training taken over by the KTH in 1869. Unfortunately, it is not possible to distinguish these engineers from the military ones, but the education of 'civil' engineers was obviously only marginal in the school's total activity (see Torstendahl, *Teknologins nytta*, p. 20).
46. See *Betänkande och förslag angående den lägre tekniska undervisningen i riket*, 21 November 1874. Printed in *Bihang till Riskdagens protokoll vid lagtima riksdagen i Stockholm, år 1876*. For Chalmers' position in the organisation of the technical education, see p. 175.
   As early as 1852, the Chalmers school had been divided into a higher and lower division, but this was never explicitly mentioned in its statutes until 1876. (See *Betänkande . . . den högre tekniska undervisningen. SOU* 1943:34, p. 46 and A. Wijkander, *Chalmerska institutet 1829-1904*, Gothenburg 1907, p. 29.) Formally, it is only at the latter date that a division in *fackskolor* (*Fach* schools) is mentioned, but such a division had actually existed for many years (see *SOU* 1943:34, p. 46).
47. T. Lundborg, 'En blick på tekniska högskolans historia', *Teknisk Tidskrift*, 19 Sept. 1927, p. 343. Also in T. Althin, *KTH 1912-62. Kungl. Tekniska Högskolan i Stockholm under 50 år*, Uppsala 1970, p. 12.
48. A comparison with the polytechnical institute in Vienna is interesting at this point. The statutes of the Vienna school were ten years older than the Stockholm ones, and the latter resembled them in several ways. The Vienna statutes stated explicitly that the main task of the institute was to provide a *scientific* education. By so doing it forestalled criticism from the crafts guild, which might otherwise have recognised a threat against, and competition with their privileges. ('Scientific' was italicised in the statutes. See Henriques, 1917, pp. 17, 101.) In Stockholm exactly what was foreseen in Vienna happened. While the *fabriks- och manufaktursociteterna* (factory and manufacturing societies)

considered the establishment of a technical institute very useful to Swedish trade and industry, the *hantverk societeten* (crafts society) was against the idea in that respect. Henriques has emphasised how the view of the crafts society was due to a misjudgement of the purpose of the suggested educational institution (p. 93). It should be pointed out, too, that the polytechnical institute in Prague – founded in 1806 – did not admit sons of craftspeople if they intended to run the crafts in the future. According to Henriques, it was thus established that the purpose of the school was other than merely to improve pure craft skills (p. 94).

49. See, for example, J. Guinchard, *Sweden. Historical and Statistical Handbook*, Stockholm 1914, Vol. I, p. 417.

50. See Henriques, 1917, p. 219. It was also stated that the school, whenever called upon to do so by the governmental departments, had to comment on subjects concerning industry/crafts and to supply craftsmen in the private sector with advice and information. This function as an advisory institution had been one part of the institute's double aim since its foundation.

51. 'Scientific education' is italicised here.

52. Estimated from *Ingénieurs Diplômés* 58/59, October 1974, Table 1. The schools are collectively called the *Grandes Écoles*, but when this designation is used we think of the more prestigious ones, especially in the *École Polytechnique* and its *Écoles d'Application* as well as, for example, *École Centrale des Arts et Manufactures*. J. Ardagh in *The New French Revolution. A Social & Economic Survey of France 1945-1967*, London 1968, talks about 'The dozen or so great colleges known as the Grandes Écoles' (p. 332) and in a study 'Le Prix des Cadres', *L'Expansion*, June 1976, the expression 'Grandes écoles d'ingénieurs' referred to the following nine schools only: *Polytechnique, Centrale* (Paris), *Supélec* (*École Supérieure d'Electricité*), *Télécommunications, Ponts et Chaussées, Sup Aero, Mines* (Paris, Nancy, Saint-Étienne) (p. 184).

The *Écoles des Mines* in Saint-Étienne and Nancy was founded in 1816 and 1918 respectively.

53. Total 1974, 10 150 graduates.

54. A calculated figure for 1955 according to the method used here stated the number of diplomas that year as being about 7 800, while the actual number of diplomas was 4 158 (see *École Centrale Annuaire* 1976, p. 53). This means that the true number was only 53 per cent of the calculated figure.

We know that the actual number of diplomas at *École Polytechnique, Mines* (Paris) and *Centrale* in 1900 and 1950 was roughly the same, but less than 50 per cent of the 1974 number (*Mines*), two-thirds (*Polytechnique*) or about four-fifths (*Centrale*).

Based on the test of our calculations for 1955, it seems reasonable to reduce the calculated figure for 1900 by about 50 per cent. Around 1850 the number of graduates at *École Polytechnique* was one third to one half of the number for the 1970s, at *Mines* and *Centrale* about a quarter. Consequently a reduction by about 75 per cent seems reasonable.

55. Even these figures could be too high. By way of a diagram M. Levy-Leboyer, in his article 'Le patronat français a-t-il été malthusien?', *Le Mouvement Social*, July-September 1974, nr 88, presents information about *Polytechnique, Mines* (Paris and St. Étienne) and *Centrale* (p. 23). If the total number for these schools in 1900 is doubled, we obtain approximately 1 000.

56. Calculations for the eighteenth century are based on information about *École Polytechnique, École des Mines* (Paris), *École des Ponts et Chaussées* and *École du Génie Maritime*. As approximate average for the annual number of graduates since the schools were founded, the following figures are used: 10 *Mines*, 25 *Chaussées*, and for *Maritime* 10 per year. Assuming that the death rate and total active period agree with the figures stated below, the total number of

engineers in the society around 1800 with education from these schools will be about 1 250.

The average number of graduates during the period 1801-1820 (*Polytechnique, Chaussées* and *Maritime*) is estimated as being 175 per year, during 1821-30 (these schools as well as *Mines* – Paris and St. Étienne) 200, and during 1831-50 (these schools and *École Centrale*) 275 per year.

The total active period of the engineers is assumed to be 40 years. The death rate is based on the averages for Swedish death rates during 1801-1850 and 1851-1900 for age groups 20/24-55/59 years. We here obtain an average death rate of approximately 17 per thousand, which implies halving an age group after 40 years. (Data from *Historisk statistik för Sverige*, Part I, Population, Stockholm 1969, Table 40.)

This figure being based on a total view of available information concerning the condition of the male population in Sweden, it entails an overestimation of the conditions in France, particularly for these occupational categories.

The result of the calculations produces a total number of 6 500 around 1850. For the period 1850-1900, the annual number of graduates has been assumed to increase from 275 to 1 000, which corresponds to an annual increase of 2.6 per cent. The same rate of growth has been used for the succeeding period up to the War. Here, too, the death rate is taken to be 17 per thousand and the active period 40 years.

57. Statistics on population from B.R. Mitchell, *European Historical Statistics 1750-1970*, London 1975, Table C1.

58. Statistics from Manegold, 1970, pp. 320-1, up to the turn of the century, and for the following years, Pfetsch, 1972, p. 186. It should be noted that Pfetsch's figures are lower than those given by Manegold.

59. Calculated according to the same assumptions concerning death rate and the active period as in the case of France. (See note 56.) Here it should also be pointed out that the death rate is an overestimation of the true conditions of this professional category.

All students in the statistics for the nineteenth century are assumed to have finished their education and entered the society three years after the observation of the total numbers in attendance at the *Technische Hochschulen*. This entails a source of error when calculating, as the duration of study for certain types of education at the *Hochschulen* was extended to about three-and-a-half or four years during the latter part of the century.

However, the duration of study at *Abteilung für Maschinenwesen*, for example, at Karlsruhe TH was only about three-and-a-half years and at *Abteilung für Chemie* three years in 1893. (See *Statistik der Grossherzoglich Badischen Technischen Hochschule zu Karlsruhe. Weltausssstellung zu Chicago 1893*.) At *Abteilung für Maschinen-Ingenieurswesen* at Berlin TH, the duration of study in the middle of the 1890s was calculated to be only three years. (See *Chronik der Königlichen Technischen Hochschulen zu Berlin 1884-1899*, Berlin 1899, p. 183.)

In the process of assessing the number of engineers in the period 1900/04, the duration of studies has been assumed to be three-and-a-half years; for the rest of the studied period, it has been taken to be four years.

It should be pointed out here that our calculations show a greater number of well qualified engineers in the society around 1900 than was accounted for by K-H. Ludwig in *Technik und Ingenieure im Dritten Reich*, Düsseldorf 1974.

He writes: 'Aufgrund einer groben Schätzung dürfte es um die Jahrhundert-wende in Deutschland rund 30 000 akademische Ingenieure gegeben haben und etwa drei-bis fünfmal so viel Fachschulingenieure sowie als Ingenieure tätige Autodidakten. Die Gesamtzahl der "Ingenieure" belief sich damals demzufolge auf rund 150 000' (p. 19, note 6).

Unfortunately the method of estimation is not mentioned.
In another context in his work, however, Ludwig would support the calculations made in the present discussion. He mentions that the figure of the *VDI* members just before the War was close to 25 000 and that this figure made up more than 10 per cent of the total number of German engineers (p. 26). If we assume the same relative proportion of highly qualified engineers before World War I as around 1900, the calculations made here would appear to be fairly reasonable.

60. No statistics are available concerning the economically active population of Germany in earlier periods.

61. See note 45.

62. The average annual number of graduates at Marieberg from 1820 was about 25. (See Sylvan, Kuylenstierna, 1918, p. 41 ff.) The average figures for Chalmers were 12 during the 1830s and 18 during the 1840s. (See Bodman, 1929. Diagram at p. 200.) The KTH number has been assumed to be of the same relative proportion as the number of the technically educated at Chalmers and KTH 1850-80 – where the figure for Chalmers made up 86.6 per cent of the KTH number. (See Torstendahl, *Dispersion of Engineers*, p. 43.) In all cases, the death rate has been put at 17 per thousand.

63. 700 from KTH, 500 from Chalmers and 600 from Marieberg. The calculated figures for KTH and Chalmers are based on the total number of students of all categories and an assumed average two-year duration of studies. Death rate: 17 per thousand.
Statistics from Henriques (KTH), 1917 and C. Palmstedt, Ed. v. Schoultz, *Historisk öfversigt af Chalmerska stiftelsens och statens teknologiska läroanstalts tillkomst och undervisningsarbeten m.m.*, Göteborg 1869, Bilaga Litt. A. From 1850 onwards the calculations are based on figures of graduates given in Henriques and Bodman.

64. According to the 1907 public investigation concerning Swedish lower technical education, the total number of employed technicians in Sweden 1908 with an education from KTH and Chalmers was 1477 (see Table 2.7), which is only 50 per cent of our calculated figure. However, as was emphasised in the investigation, the committee did not aspire to give the number of all technicians working in Swedish industry in 1908, that is, technicians graduated from technical schools. The stated figure only covers those technicians who were *employed*, and the investigation did not consider the 'surely considerable number' of consulting engineers. The information about architects and building contractors was considered incomplete, 'as a majority of them run their own business' (p. 486). The members of *Kungl. väg- och vattenbyggnadskåren* (Royal Corps of civil engineers), graduates of the KTH, were also excluded from the investigation. Information concerning, for example, the members of *Svenska Teknologföreningen, STF*, made it possible for the committee to check its result. Of the total number of members for 1908, the investigation found that 993 were persons with such an employment 'that they should have been included in the committee's statistics' (p. 486). This figure was very close to the figure of those graduated from KTH, a figure which the committee had studied (972 persons). However, it should be noted that the figure 993 is only about 55 per cent of the total number of *STF* members, in 1909 (see Table 2.12).
Although our calculations might have supplied rather too high figures, they probably give a fairly accurate picture of the true situation concerning the *total* number of practising qualified engineers in the beginning of the twentieth century. The number of *STF* members in 1909 made up about 60 per cent of our calculated figure that year, which seems to be a realistic member rate.
With regard to an earlier period, it is possible to check the relevance of our

calculations. The register of KTH graduates 1850-80 — a total of 803 graduates — has been studied by Rystedt (See C.G. Rystedt, 'Teknologernas verksamhet och öden', *Industritidningen Norden*, 1881, pp. 39, 44-6.) He found that at the end of the period 63 of the 803 graduates were dead, which left 740 active ones. According to our calculations, the number of practising KTH engineers at the end of 1880 was 775, which must be considered an acceptable figure as our calculations are based on the assumption of an active period of 40 years.

65. As far as we know, no directly comparable information is available concerning the structure of industry in France and Germany during the nineteenth century. Nor is there any information available on the proportions of sectors in private and public control to enable a comparison.

Nevertheless, there is no reason to believe that, in the latter case, the conditions in France were essentially different from those in the German states and in Germany, as governments were responsible for the infrastructure here, too — roads, bridges, canals, railways, postal services — and owned, for instance, several mines, foundries and factories. See W. Fischer, 'Government Activity and Industrialization in Germany (1815-1870)', in W.W. Rostow (ed.), *The Economics of Take-Off into Sustained Growth*, London 1963, chapter 5.

66. One way of establishing in which industrial sectors the former students were practising might have been to study the students' choice of educational specialisation. The table below shows this specialisation during the 1870s and during the first decade of the twentieth century.

	1870s		1900s	
	Number	%	Number	%
Mechanical:	68	43.3	627	48.4
Metallurgists:	27	17.2	286	22.1
Production/Design:[a]	45	28.7	298	23.0
Chemists:	17	10.8	84	6.5
	157	100.0	1295	100.0

Note:
[a]*Constructeurs.*
Source: Comberousse, 1879, p. 212 and M.E. Coignet, *Rapport présenté au conseil de l'École Centrale. Au nom de la Commission spéciale chargée d'examiner la question dite 'de la Spécialisation'*, Paris 1910, p. 9.

However, it was found that for the 1870s more than 60 per cent of the graduates chose other careers than their specialisation indicated (Comberousse, 1879, p. 213) and of the 1 295 graduates during the first decade of this century only 513, i.e. 40 per cent, stayed in the field of specialisation they had chosen at the school. Consequently, this is a very unprecise way of finding out where the graduates of the *École Centrale* were practising.

67. Artz, 1966, p. 252.
68. Comberousse, 1879, p. 250.
69. Levy-Leboyer, 1974, p. 22.
70. Comberousse, 1879, p. 251.
71. France's partners obtained access to the French market and tariff rates were limited to a maximum of 25 per cent.
72. T. Kemp, *Industrialization in Nineteenth-Century Europe*, London 1969, p. 69.
73. Ibid.
74. A total of 841 former graduates of the school were, in different functions, active at the exhibition; 727 as exhibitors under their own names or together with others; 340 won medals or received diplomas, decorations and mentions — 92 gold,

97 silver and 68 bronze medals. According to Comberousse, probably no other group of exhibitors with the same origin could present a more remarkable result. See Comberousse, 1879, p. 257.

75. *Report of the Schools Inquiry Commissioners on Technical Education.* Parliamentary Papers, 1867, Vol. 26, p. 267.

It should be pointed out that Playfair in his 1852 study – *Industrial Instruction on the Continent* – already praised *École Centrale*, and spoke of its 'immediate and eminent success' and as 'now the most important industrial institution in France' (p. 26).

By citing a contemporary report, Report of the Commission of the Chamber of Deputies, in order to inquire into the school's budget, Playfair emphasised *École Centrale*'s complete success: 'This is confirmed both by the unanimous opinion of the first manufacturers of the country and by the ease with which all the pupils educated at it have received employment' (p. 27). He also stressed how 'a certificate from this institution is equivalent to assured success in life. Its pupils invariably pass into the most important positions in industry; and not only France, but Spain, Belgium, and England have learned to value them, as we see by the ready manner in which the manufacturers of these countries secure their services'.

76. Levy-Leboyer, 1974, p. 22.

77. Comberousse, 1879, p. 250. The precise figures were 7 266 and 4 054. Of the latter figure, 552 were foreigners (p. 212).

78. Artz, 1966, p. 239.

79. Ibid.

80. Ibid.

81. Our calculations here are based on the assumption that all admitted students also graduated. Although we know that very few of the students failed, this entails a (very) small overestimation. However, this overestimation is counterweighted by another error in the calculations. A student admitted in, for example, 1833 is supposed to graduate in 1835 and without taking further training at an *École d'Application*.

The result is very much in line with Artz's results for *École Polytechnique*'s first ten years. Artz says that 'by 1806, of the 1664 students admitted, about a thousand held various state positions' (p. 159). In percentage terms this means that 60 per cent were in governmental services. Most of the rest are said to have entered commerce and manufacturing.

Still, it should be noted that according to Levy-Leboyer, only 80 polytechnicians or 1.09 per cent of the total number who left the school had taken up services within industry or commerce during the first half of the nineteenth century. See Levy-Leboyer, 1974, p. 24.

82. Artz, 1966, p. 247.

83. M. Aguillon, *Supplément à la Notice Historique sur l'École Nationale Supérieure des Mines*, Paris 1899, p. 6.

84. See *Les Écoles Nationales Supérieurs des Mines*, Nr 15, November 1961, p. 297.

85. See Artz, 1966, p. 246.

86. Levy-Leboyer, 1974, p. 22, footnote 23. Exact figures for the period 1850-1914 (five-year averages) were:

1850-54	108	1870-74	197	1890-94	251
1855-59	137	1875-79	216	1895-99	292
1860-64	173	1880-84	246	1900-04	314
1865-69	198	1885-89	249	1905-09	306
				1910-14	351

87. Calculated according to the assumptions of an active period of 40 years and a death rate of 17 per thousand. See note 56.

88. Unfortunately, as far as we know, there is no information concerning the graduates' choice of careers, of the kind available for certain French schools.

89. P. Lundgreen, *Techniker in Preussen während der frühen Industrialisierung. Ausbildung und Berufsfeld einer entstehenden sozialen Gruppe*, Berlin 1975, pp. 226, 276.

90. Ibid., pp. 219, 243 ff, 273 ff.

91. Ibid., pp. 277-8.

92. Playfair, 1852, p. 15.

93. Ibid., p. 25.

About the Polytechnic School of Munich, he wrote that 'it is undoubted that its pupils are in great demand, and fill important positions in industry' (p. 22).

Referring to people like Redtenbacher, he pointed out concerning the Polytechnic School at Vienna, that 'notwithstanding the large number of students, the demand for them by industrial establishment is greater than can be readily supplied' (p. 20).

94. Landes, 1972, p. 347.

95. Pfetsch, 1974, p. 179.

It should be noted that while he said that the contribution of science and technology in the growth process was large and growing, he also stressed that the causal relationship is dubious, as well as the extent of its contribution (p. 129 ff).

In this context Pfetsch refers, among other things, to a study by D. André, *Indikatoren des technischen Fortschritts* (Göttingen 1971), for Germany 1850-1912, which showed that the residual factor contributed 42 per cent of the total economic growth, and 38 per cent within the industrial sector alone. In the study the technological factor was divided into two components, one structural and one educational, and according to André the improved quality of labour contributed to almost one third of the residual (31 per cent in total, 32 per cent in the secondary sector alone).

96. Pfetsch's definition of *Wissenschaftspolitik* is very general:

Unter Wissenschaftspolitik sollen alle Massnahmen verstanden werden, die auf Lehre und Forschung in Hochschulen, ausseruniversitären Wissenschaftseinrichtungen und Forschungs- und Versuchs-laboratorien der Wirtschaft gerichtet sind und von privaten Personen, Gruppenorganisationen und staatlichen Organen getragen werden. Das direkte oder indirekte, beabsichtigte oder unbeabsichtigte Ziel solcher Massnahmen ist, die Erweiterung des Wissens zu ermöglichen (pp. 30-1).

97. Pfetsch, 1974, p. 139.

98. Ibid., p. 179.

99. Pfetsch, 1974, chapter four 2.2., 2.3.1. and 2.3.2., pp. 166-81. The great number of basic inventions during the latter part of the nineteenth century, as well as the continuously decreasing number of years between the time of the invention and its commercial use, should also be seen as a part of this process.

100. H. Schimank in *Der Ingenieur. Entwicklungsweg eines Berufes bis Ende des 19. Jahrhunderts*, Köln 1961 has stated that the development of the *Maschinen- und Fabrikswesens* — which he sees as a result of the work of a small group (*Kreises*) of very intelligent technicians — 'wurde bald zur Ursache einer ständig wachsenden Nachfrage nach Ingenieuren' (p. 38).

Schimank discusses the engineer in general. But, bearing in mind the large number of those with an education from a technical university — to which should be added engineers with lower qualifications — the demand for engineers was probably satisfied.

Concerning engineers in management and leading positions, a 1912 study – F. v. Handorff, 'Die Verwendung der Hochschulsabsolventen im Staatsdienst, in der städtischen Werken und Verwaltungen und in der Industrie' in *Abhandlungen und Berichte über technisches Schulwesen*, Berlin 1912, Band IV – emphasised that without any doubt 'heute die Mehrzahl unserer industriellen Unternehmungen von technisch gebildeten Direktoren geleitet wird' (p. 85). This was especially the case in *Bergwerks- und Hüttenindustrie*.

101. *Betänkande och förslag till utvidgning och omorganisation af Tekniska Högskolan*, Stockholm 1891, p. 70.

102. See J. Wallander, 'Ingenjörerna i studentbetygen och i verkligheten', *Teknisk Tidskrift* 1944, p. 988. It is a case study of students who finished their secondary school (*studentexamen*) during the years 1881-1909 and went into technical occupations. A total of 233 persons were studied, and 95 per cent of these had studied and/or graduated at the KTH.

103. The Committee of 1919 for the reorganisation of Chalmers. Here after *SOU* 1935:52 (*Bet. . . . tillströmningen till de intellektuella yrkena*) pp. 258-62. In connection with the appointment of the Committee of 1919, the head of the ministry said that within professional circles it has 'for a long time' been emphasised that fewer engineers than are required to cover the need – 'not only of the present industry, but especially in order to create suitable conditions for the new industry' – are educated. The first 'parent' of this opinion was said to be *Svenska Teknologföreningen*. In an address to the government *Sveriges Industriförbund* had expressed the same views (p. 263).

104. See note 64.

Information also exists concerning engineers graduated from the KTH 1890-1891-1892 (162 persons). See Henriques, 1927, p. 190.

105. Besides the main investigation, a subsidiary study was also made, engineers qualified at KTH between 1860 and 1868 (206 persons) being among the objects of study. No useful figures regarding Chalmers were obtained for this decade. See Torstendahl, *Disperson*, etc., pp. 77-8.

106. See for example Kemp, 1969, p. 102.

107. *The Cambridge Economic History of Europe*, Vol. VII:1, p. 517. Kocka dates the German industrial revolution to the middle third of the nineteenth century.

108. W. Fischer, *Der Staat und die Anfänge der Industrialiserung in Baden 1800-1850*, Band I, 'Die staatliche Gewerbepolitik', Berlin 1962.

109. Lundgreen, 1975.

110. Fischer, 1962, p. 163.

111. Lundgreen, 1975, pp. 119-20.

112. Ibid., p. 119.

113. See K. Karmarsch, 'Die Polytechnische Schule zu Hannover' (1856), in W. Treue, H. Pönicke, K-H. Manegold, *Quellen zur Geschichte der industriellen Revolution*, Göttingen 1966, p. 122.

114. Landes, 1972, pp. 346-7.

115. For example, see Manegold, 1970, p. 75.

That the engineer possessed a 'hoch entwickeltes soziales Selbstbewusstsein' has also been stressed; see H. Klages, G. Hortleder, 'Gesellschaftsbild und soziales Selbstverständnis des Ingenieurs', *Schmollers Jahrbuch*, 85:6 (1965), p. 670.

116. Manegold, 1970, pp. 75-6.

See also Runeby, 1976, chapter IV, in this context and Ludwig, 1974, p. 22.

117. Manegold, 1970, p. 83.

118. Treue, 1964, p. 235.

119. G. Goldbeck, *Technik als geistige Bewegung in den Anfängen des deutschen Industriestaates*, Berlin 1934, p. 28.

120. Ibid., p. 31.
121. See pp. 23-4 and, for example, Runeby, 1976, chs. 3 and 4.
122. Eriksson, 1978, p. 73.
123. Runeby, 1976, p. 180.
124. For example, the number of admitted students 1850/54 was 90 per cent, but two out of three were admitted 1855/59 and in 1860/64 78 per cent. The calculated figures supplied here are based on statistics from *Bet. . . . Tekniska Högskolan*, pp. 60-1, Table 2; and *Bet. . . . intellektuella yrkena, SOU* 1935:52, p. 242. Table 74 (1880s-1914).
Certain calculations for the years 1848-96 are also provided in V. Adler, *Om det tekniska undervisningsväsendet i Sverige*, Stockholm 1897, p. 57.
125. In the public investigation on Swedish higher technical education – in the context of KTH and the period 1825-1911 – we read: 'it is striking how precisely the business cycles are reflected in the existing tables concerning the number of applicants'. *Bet. . . . den högre tekniska undervisningen, SOU* 1943:34, p. 20.
126. Calculated from *SOU* 1935:52, p. 262, Table 86.
127. Table in *Undervisningsväsendet. Statens Allmänna Läroverk för Gossar. BiSOS* 1897.
128. See P. Dahn, *Studier rörande den studerande ungdomens geografiska och sociala härkomst*, Lund 1936.
129. A strong tendency for the student to choose the same or similar profession as his father was also indicated. See Dahn, 1936, p. 386. Torstendahl in his study *Dispersion . . .* has also established a strong relation – 'no illusionary causal relation' (p. 259) –between a father's occupation as a business leader in the private manufacturing industry and the son's corresponding position 30 years after graduation. To quote Torstendahl: 'More than ½ of the manufacturing industry business owners with a KTH or Chalmers education had fathers who were business leaders; . . . Of employed business leaders in manufacturing industry companies with up to 100 employees, somewhat less than ½ had fathers who were business leaders themselves, while of the employed business leaders of manufacturing companies with over 100 employees more than ½, for the KTH subjects as much as two thirds, had fathers who were business leaders' (p. 255). Torstendahl's main study concerned the graduates 1880-1910. Consequently the study is to a certain extent relevant to our period, too.
130. Refers to 1976 for *École Central* and 1972 for *Ecole des Mines* (Paris). The calculation for *École des Mines* is based on information in *École des Mines* (Paris, St. Étienne). *Armines, Rapport d'activité* 1972, p. 20 and *ENS des Mines de Paris. Cycle de Formation des Ingénieurs Civil*, p. 19, whereas the information for *École Centrale* is based on statistics from the school archive.
131. See Artz, 1966, p. 23.
132. Calculations based on statistics on applicants and admitted students in *Repertoire de l'École Imperiale Polytechnique*, Paris 1855 and 1867, Table c and pp. 45-50, respectively; J.P. Callot, *Histoire de l'École Polytechnique*, Paris 1958, p. 348; and M. Hachette, *Correspondance sur l'École Imp. Pol.*, Paris 1813, p. 128.
133. Comberousse, 1879, p. 209. Statistics also provided in L. Guillet, *Cent Ans de la Vie de l'École Centrale des Arts et Manufactures*, 1829-1929, Paris 1929, p. 77.
134. Calculations from Guillet, 1929.
135. P. Laffitte, *Les Écoles d'Ingénieurs en France. La documentation française*, 3 December 1973, Nos. 4045-4046-4047, p. 8.
136. Ibid., p. 37.
137. These days, some schools have grouped together to organise the entrance examinations, and in the event of success by the students at the *concours*, it will

give access to a certain number of schools.

That the *École Polytechnique* is on top of the ladder of prestige is obvious, as almost none of those accepted to the *École* chose to study at another school.

In the *concours Mines-Ponts*, common for seven schools, the preference is for *École des Mines* in Paris, then *École des Ponts et Chaussées*, etc. In the *concours Centrale-Supélec*, common for four schools, the first choice is *École Centrale des Arts et Manufactures*.

The *concours* follow a period (2-3 years) of compulsory post-secondary schools studies at specific *classes préparatoires* at the schools. Only students with a first-class *baccalauréat* are accepted in these forms and in, for example, the academic year 1969-70 it was found that 36 per cent of the students with this examination in science the preceding year were found in the preparatory classes (see Laffitte, 1973, p. 39). Thus a selection of the students is made at two levels, first after the *baccalauréat* at the entry to the preparatory courses, and second at the *concours* which after successful results gives admission to the school.

138. Laffitte, 1973, p. 38.

139. Ibid.

The generally prestigious engineering education also seems to be reflected in the French salary structure, as well as the ranking order of the *Grandes Écoles*.

Thus, according to the above-mentioned study in *L'Expansion* (see note 52), it was found that the highest salaries in French society in the middle of the 1970s are paid to graduates from the *Grandes Écoles* with *École Polytechnique* on top, followed by *École des Mines* and *École Centrale*.

140. See Henriques, 1917, p. 8.

141. Kindleberger, 1976, p. 5.

142. See his 'French Entrepreneurship and Industrial Growth in the Nineteenth Century', *The Journal of Economic History*, Vol. IX: 1 (1949), pp. 45-52.

143. Landes, 1949, p. 55.

144. Ibid., p. 56.

145. Ibid.

146. Ibid., p. 57.

147. Landes writes: 'For the purpose of this article, the entrepreneur is not only the "innovator" as such but the adapter and manager as well. In other words, the entrepreneur is the businessman who makes the decisions' (p. 46, footnote 3).

148. Kindleberger, 1964, p. 158.

149. Ibid., p. 185.

150. The border-line has been put to the defeat of the Freycinet Plan in 1883 and its subsequently reduced scope. The original plan of 1879 was intended to meet the recession of the late 1870s with a public programme of extending railroads, canals and roads.

151. Kindleberger, 1974, p. 185. The only exception during the six decades was the rebuilding of the French eastern region after World War I. No positive governmental intervention to promote economic growth, of proportions similar to those during the Second Empire, was taken until the middle of the 1940s.

152. Examples in this context in the nineteenth century are shipbuilding, chemicals (synthetic dyes, soda), aluminium (Allais-Camargues), glass (St Gobain), automobiles (Citroën) and locomotives (the steel companies, Five-Lille and Gouin); see Kindleberger, 1974, p. 156, note 87.

153. Kindleberger, 1974, pp. 157-8.

154. Ibid., p. 158. Fohlen has emphasised the growing status of technicians in nineteenth century France. See his 'Entrepreneurship and Management in France in the Nineteenth Century' in *The Cambridge Economic History Europe*, Vol. VII:1, p. 373 ff.

155. D.S. Landes, 'French Business and the Businessman: A Social and Cultural Analaysis', in E.M. Earle (ed.), *Modern France. Problems of the Third and Fourth Republics*, Princeton 1951, p. 349.

156. Gerschenkron, 1965, p. 70.

157. Ibid., p. 65.

158. See Landes, 1972, p. 529.

159. Kindleberger has emphasised the last fact, but also said that it is not to claim that French technical education was admirably suited for success in business; 'its Cartesian quality, emphasis on mathematics, and elitism produced a succession of brilliant and arrogant technocrats who did best in the army and public works and less well, unless they had apprenticed abroad, in the family firm.' See his 'Germany's Overtaking of England 1806-1914', *Weltwirtschaftliches Archiv*, Bd 111:3 (1975), p. 503. Also printed in *Economic Response. Comparative Studies in Trade, Finance, and Growth*, Cambridge, Mass., and London 1978.

160. In contrast to in Britain, where it was approved *by* the elite, Kindleberger, ibid., p. 503.

3 ENGLAND

3.1 Higher Technical Education – The Exception

Scientific education for occupations within industry was for a long time lacking in England and the only technical education of a more extensive character during the early half of the nineteenth century was the kind offered at the Mechanics' Institutes, a large number of training centres founded during the 1820s.

However, the activity at these institutes was on a lower level, technical as well as social, and, to quote Mathias: 'In no sense were they the context from which might spring a technical university system, as in continental countries.' In addition, the mechanics' institute movement petered out after the middle of the century.[1]

However, there were certain noticeable occurrences around 1850. The Great Exhibition in London in 1851 was to some extent the starting-point for scientific education in England, as well as a point of departure in the modern history of English manufactures and arts.[2] The Exhibition thus taught English producers 'the useful lesson that England possessed no monopoly of inventive genius or practical skill', along with the fact that foreign producers had advanced further in aspects of industry which required scientific application.[3]

An immediate result of the Exhibition was the founding of the School of Mines – from 1863 the Royal School of Mines – and the Department of Science (1853) under the Board of Trade. The new department also took over the Royal College of Chemistry, founded in 1845.[4]

The enlargement and improvements of the English technical education system went slowly, however. Published in 1869, J. Scott Russel's extensive work *Systematic Technical Education for the English People* – which considered technical education at all levels – established that in England 'As a rule, technical education does not exist.'[5] Obviously, the real alarm signal was the great Paris Exhibition in 1867. Scott Russel – who acted as a juror at the exhibition – expressed the English experience in the following words: 'By that Exhibition we were rudely awakened and thoroughly alarmed. We then learnt, not that we were equalled, but that we were beaten – not on some points, but by some nation or other on nearly all those points on which we had prided ourselves.'[6] On the ninety classes there were 'scarcely a dozen' in which pre-eminence was 'unhesitatingly awarded' to England.[7]

One immediate result of this exhibition was the technical instruction movement;[8] others were several public investigations in these matters.[9] Recommendations were only realised to a minor extent, though.[10]

The Great Depression and a greater awareness of a growing international competition induced changes in the English educational system. Of particular interest for the higher technical education was the Central Institution at South Kensington, founded by the City and Guilds Institute in 1884. From 1893, it was called the Central Technical College; in 1901 it became a School of the University of London, and 1907 part of the Imperial College — which has been characterised as 'one of the first places in the country to teach applied science at a high level',[11] and as an English counterpart to a German or Swiss polytechnic.[12] The City and Guilds Institute also founded technical colleges from the 1880s onwards. The Finsbury Technical College was the first of them (1881); the training they offered was at a slightly lower level. Cotgrove has emphasised that the years 1880-1905 formed a period of rapid growth in further technical education,[13] and that the importance of increased state aid was realised.[14]

England's first Technical Instruction Act was passed in 1889. According to Cardwell, this meant that the battle for state-aided technical education was won; 'the scene was now quite modern with polytechnics, technical colleges, etc. The Mechanics' Institutes were no longer a living issue'.[15]

However it should be emphasised that in comparison with the development on the Continent, especially in Germany, England still lagged behind, quantitatively as well as qualitatively. For example, it has been pointed out that the English polytechnics were not colleges in the German sense, 'but an English blend of philanthropy, adult education, and technical instruction'.[16] At the beginning of World War I, the English system still had 'a long way to go to catch up with the German — at least from the standpoint of economic productivity'.[17]

In a sense, it would be fairly easy to illustrate the factual development of the higher technical education in England during the nineteenth century, since only a small number of English schools during that century actually provided an education that corresponded to international standards. However, statistics as to the total number of students would not seem to be available. Instead, some scattered information concerning the number of students at the above mentioned institutions of higher technical education — the College of Chemistry, the School of Mines and the Central Institution – will illustrate the limited extent of that education.

During its initial 25 years, the number of matriculated students at the Royal College of Chemistry was fairly constant, ranging from 40 to 50,[18] while the annual average at the School of Mines during the 1850s was a mere 14 full-time students. The number of part-time students at the latter school during this decade was about three-and-a-half times the number of full-time students.[19] To the Samuelson Committee (1868) it was said, among other things, that the number of students at the School of Mines the previous year was only 18. The Committee also made the important point that 'Generally this small school was not full.' The reason was attributed to 'public indifference to science'.[20]

In 1886/87 the Central Institution at South Kensington had about 640 students; less than 50 of them were doing the complete three-year course, though. A few years into the 1890s, the latter figure had quadrupled.[21]

Towards the end of the century, there was a growing number of students taking courses in science and higher technology at the English universities (Cambridge, for instance, received a Chair in engineering in 1875[22]), at the schools mentioned in the present discussion, and at other university colleges[23] and colleges 'with some pretensions to university standards, and the principles of scientific and technological teaching and research'.[24] In spite of this increase, the total number of students in these subjects in 1899 has been estimated to approximately 2 000 only.[25]

In their recent study on the subject Roderick and Stephens have concluded that 'it would appear that the number of fully-trained scientists and technologists in England fell far short of what was desirable in the last quarter of the nineteenth century'.[26]

Further increases in the number of students in these subjects took place during the first decade of the twentieth century, and according to information regarding the year 1912/13, the total number of full-time students at universities, colleges, and technical institutes in England and Wales was approximately 2 700. Fifty-five per cent of them studied at universities and colleges and the rest at technical institutes.[27] However, the number of graduates in 1913 in *all* branches of science, technology, and mathematics with first- and second-class honours in England and Wales was only 350.[28]

The Imperial College – founded by royal charter in 1907 out of the

Royal College of Science,[29] the Royal School of Mines and the Central Technical College, and admitted as a school of the University of London in 1908 — was of cardinal importance to the higher technical education in England from the very first. During its first year, the number of full-time students amounted to 600, and at the time of World War I the number had increased to 800, 'all reading science and technology at an advanced level'.[30] This means that the English higher technical and scientific education was very concentrated, and that probably more than 50 per cent of all students at universities or colleges in 1914 were studying at the Imperial College.

The statistics are to some degree comparable to those that apply to the technical universities in Germany. However, as has been indicated, the English technical education was both less systematic and less advanced.[31]

An important aspect in this context is the fact that the higher technical education to a large extent took place at the universities and not at separate institutions. This was probably — in various ways — unfortunate for the English industrial development. On this point, I disagree with the view put forward by Sanderson in his work *The Universities and British Industry 1850-1970*. Sanderson feels that a changed attitude to the education of the engineer could be discerned towards the end of the 1860s and in the beginning of the following decade — he talks about 'a watershed in this regard' — and that the university came to play a more important role. Faced with two distinct traditions — the English, which lay stress on apprenticeship outside the university, and the Continental tradition 'which believed it possible to give virtually the whole of engineering education within colleges such as those in Zurich, Stuttgart, and Karlsruhe . . .', contemporary decision-makers chose 'a middle way — university followed by apprenticeship work . . .'[32] According to Sanderson, Fleeming Jenkin advocated this middle way in his Inaugural as the first Professor of Engineering at Edinburgh in 1868; 'This wise and clear statement . . . set the pattern of the British approach to higher engineering training to this day.'[33]

Oxford, by and large, was unable, or unwilling, to adjust to industry's new requirements, and Cambridge disengaged itself from the industrial purpose (its science was, however, of a very high international standard). Hence, a new form of higher education had to be created in the North and the Midlands: the new colleges or, as they have been called, the Civic Universities. Sanderson argues that 'The example of the success of this alternative tradition was perhaps the most important influence aiding change in Oxford and Cambridge after 1900, just as the defects of the ancient universities before the 1890s had been a powerful stimulus to the creation of the civic universities themselves.'[34]

Undoubtedly these new universities were important to England's industrial development. However, the essential point is that the results before 1914 would have been much better if a Continental type of education had been chosen, something that several critics of the English system proposed (see p. 15). Consequently, I feel that Sanderson — after summarising the contributions of the Civic universities — goes astray in his comparison between the English and Continental systems. He writes: 'England was saved from the dual form of higher education such as existed in Germany and which many politicians of science wished to emulate. It thus preserved the mutually beneficial linkages of science and technologies in the same institution, prevented the creation of non-technological "universities" which would rapidly have become narrow arts teacher training colleges, and so avoided a situation whereby a "second rate" stigma would have attached to centres dealing purely with technologies.'[35]

However, the successful German industrial performance and the German type of education indicate the advantages of this kind of training during the nineteenth century, a line of education that also took pre-university levels into account. Thus, the students who began their studies at the universities — classical or technical — had already acquired, at the *Realschulen* and *Gymnasien*, a broad and qualified all-round education in the humanities as well as in the pure sciences. In addition, both science and technology were taught at the Technische Hochschulen, and the 'second-rate stigma' which may have attached to the future technical universities during the first half of the nineteenth century vanished during later decades. By 1900, as we have seen, the technical universities and the classical universities were on an equal footing in Germany.

While Sanderson attributes unwarrantably favourable qualities to the English system, his account and estimation of the relevant education in Germany must be considered dubious. Indeed, it is clear that institutions such as the civic universities were in fact the ones that developed into teacher-training colleges,[36] possessing the second-rate stigma referred to above.[37]

By way of recapitulation, it is evident that scientific education was introduced comparatively late in England — at the middle of the nineteenth century — but even by the time of World War I, this education was quantitatively modest and, seen from an international point of view, of a limited scope. The dominating institution was Imperial College and its forerunners; at the latter date, more than 50 per cent of all English students in science and higher technology were probably studying at Imperial College (see above). With regard to the organisational aspect, the English higher technical education also deviated from the Continental type, as the training — apart from that offered by Imperial College — was to a large extent supplied at the universities. Considering the fact that the English educational system as a whole was less systematic and advanced, and that the type of organisation controlling the higher levels of technical education in England was an unfavourable one, the quality of the average English engineer would seem to have been lower than that of his Continental colleague.

3.2 Factors Explaining the English Development

The following discussion indicates certain factors which contribute to an explanation of the English development. These factors also suggest further shortcomings in the English system of higher technical education, shortcomings that were probably to some extent caused by the fact that separate institutions for the higher technical education were *not* founded in England.[38]

Three factors and/or indicators, strongly interrelated of course, are relevant to the English development, both with regard to quantity and quality:

(1)   The general demand for technical education, as it was manifested in industry's demand for the theoretically and practically educated engineers;
(2)   Engineering as a profession in the English society;
(3)   Society's view of activities entailing science and engineering, especially among the upper and middle classes. Here, I briefly consider the reasons why English governments during the nineteenth century failed to appreciate and introduce a unified and efficient technical education.

Discussing the development of the English technical education and its slow changes during the nineteenth and twentieth centuries, as well as the reasons for this, Landes has emphasised the great importance of the first factor; 'Essentially they [the reasons] boil down to demand, for a free society generally gets the educational system it wants, and demand was once again a function in part of British priority and German emulation.'[39] As for the future possibilities for the students in higher technical education, he summarised the English situation in the following terms: 'job and promotion opportunities for graduates in science and technology were few and unattractive'.[40] Sanderson has also established this fact, for example in his discussion on the relations between the University of London and industry. He holds that the failure of industry to absorb the output of the university was one of four main defects in their relationship.[41] However, as demand for scientifically and technically trained man-power was small, it was also 'amply met' by the existing output.[42] For example, it has been pointed out above that the number of students starting their studies at the Royal School of Mines during the 1860s was dependent on this fact, and Sanderson gives a number of similar examples concerning graduates in engineering, chemistry and physics at the beginning of the twentieth century. He concludes his discussion on this point as follows: 'The knowledge and talent were there if industry would only use it.'[43]

From a survey of the careers of 850 former students from the combined Royal College of Science and Royal School of Mines – 'the institution most likely to produce the managerial technologist and scientific research worker' – Roderick and Stephens found that only 20 per cent had entered industry at some stage in their careers, and that the majority of those who spent some time in industry entered mining and brewing and frequently held positions in the inspectorate – but not in management or research. About one third went abroad, and a slightly lower percentage entered the teaching profession.[44]

There can be little doubt as to the consequences of only employing a small number of persons with a scientific and technical education, especially in higher management (this has been shown to have been the fact in, for example, the British steel industry);[45] 'The neglect of scientific instruction in the 19th century produced generations of employers who failed to appreciate the place of scientific and technical knowledge in industry.'[46]

It thus seems reasonable to talk about a hereditary transmission of valuations, connected, among other things, with the second factor discussed here – engineering as a profession in the English society.

As was emphasised before, England's earlier industrial successes were attained by engineers who had acquired most of their knowledge by means of 'learning by doing'. Pure science was generally consulted – *via* scientifically interested gentlemen – only in cases when 'trial-and-error' or 'rule-of-thumb' methods could not solve technical problems.[47]

A development in the course of which technical activity grew into a profession started in England towards the end of the eighteenth century, but in respect to mechanical engineering – the dominating engineering category[48] – this process took a long time. In a discussion of how far this process of professionalising had gone by the end of the 1850s, Reader, in his work on the rise of the professional classes in England in the nineteenth century, argues that mechanical engineering could hardly be considered a profession,[49] while the civil engineers belonged to the professional group from the census of 1861.[50]

The growth of the engineering profession in England shows certain deviating features in comparison with previously presented characteristics. This is especially the case with regard to the comprehensiveness of the theoretical basis – i.e. the existence and use of technical institutions in the training of the engineers – but also concerning the degree of *esprit de corps*, which is reflected in the professional organisation.

The history of the English engineering organisations also goes back to the eighteenth century, when unofficial groupings without any juridical authority arose. These earlier organisations often had the character of dining clubs, as was the case with the Society of Civil Engineers founded in 1771. The civil engineers were also the first category of engineers in England to organise themselves on a national level – the Institution of Civil Engineers was established in 1818. Ten years later, this organisation received its Royal Charter. For the development of an organisation of this type in England, such a charter is important, as it gives the organisation the greatest prestige imaginable and 'may be said to confer official recognition by the State that the occupation concerned has achieved professional standing'.[51] As for the organisational development of the mechanicals, however, this was a more long-drawn-out process. A national organisation – the Institution of Mechanical Engineers – was formed as late as 1847, and not until 1930 was the organisation incorporated by Royal Charter.[52]

The fact that the mechanicals organised themselves as a separate organisation has been seen as an act of retaliation against the civil engineers, who looked down on the mechanicals. These were considered 'not strictly civil engineers, nor are they, in the sense of our definition, professional men . . .'.[53] This suggestion of snobbery affecting different English engineering organisations is a notable feature, impairing the cooperative actions initiated by these organisations in the interest of the individual engineer.[54] In order to become a member of an engineering organisation on the Continent or in Sweden, a specific theoretical-technical education was required; in England, however, the responsibility of evaluating a person's qualifications for membership had to be considered by the organisation in every single case. This was a consequence of the specifically English circumstance, referred to above, that the leading principle of a professional education was apprenticeship.[55]

By introducing written examinations for membership, arbitrariness could have been avoided; to some extent, qualifications equivalent to those operating on the Continent could have been maintained, too. However, the English engineering organisations were long reluctant to separate theoretical instruction from practical training, gauging a person's knowledge by means of such tests. The Institution of Civil Engineers did not introduce written examinations until 1897, and not until 1912 did the Institution of Mechanical Engineers do the same.[56]

On the basis of the aspects presented in this discussion, and in comparison with the Continental development, the process of professionalisation in England can be seen to present the following over-all pattern:

(a) Generally, the English process began late, especially in the field of mechanical engineering;

(b) For reasons connected with the engineers' own view of various types of engineering, a united and national organisation for the engineers, irrespective of category, was lacking;

(c) The only control of an engineer's quality was exercised by the separate engineering organisation. Written examination as a substitute for graduation from a specific technical school, and as a basis for membership in the organisation, was introduced very late.

The fact that the engineering profession developed differently in England as compared with the other countries meant that the English engineers did not for a long time — up to the 1960s — possess a united pressure group furthering their interests. In this context, there is also reason to doubt the average technical quality of, for example, the English mechanical engineer in comparison with his Continental equivalent, at least from the middle of the nineteenth century.

The factors discussed under (1) and (2), introduced with a view to explaining the English development, are also closely interrelated with the third one — the view of activities in science and engineering held by the upper and middle classes.

It has been emphasised that the social structure of England before the industrial revolution only possessed one really essential division, the one between the gentility and the common people. The crossing of this divide, in case you were born on the 'wrong' side, i.e. belonged to the common people, has been characterised as 'the great game of life to be played by anyone possessed of ambition . . .'.[57] The reasons for this aspiration have been much discussed, but it has been considered obvious that the chances of being able to cross this social divide were considerably improved towards the end of the eighteenth century and beginning of the nineteenth century; Coleman maintains that the industrial revolution was also, in a sense, a revolution for those who were not gentlemen.[58]

In addition, he argues – convincingly, judging by an investigation of the literature – that 'When the business and technological drive of the English industrial revolution loses some of its momentum in the later nineteenth century perhaps, in part at least, it is because too many of the revolutionaries are too busy becoming gentlemen.'[59]

According to English educational philosophy, the common people should receive training aimed at specific vocations, while gentlemen should be educated, a process undergone at the public schools and the classical universities. To a certain extent, pure science was included in this education – 'as an intellectual exercise'[60] – but technology was neglected; an existence characterised by living 'idly and without manual labour' (in short, the outline of a gentleman's life as it was formulated at the end of the sixteenth century, but also relevant to later centuries, at least up to the twentieth century)[61] was reflected in his education.

The fact that science was not established in the pattern of education during the scientific revolution and the early industrial revolution, and also that a reaction occurred during the nineteenth century, accentuated the distinction between 'pure' science and its application within industry; 'there evolved and sharpened in the course of the nineteenth century two parallel distinctions: between pure and applied science; and between the "educated amateur" and the "practical man". Both are social phenomena, both have continued until very recent times and perhaps are only now disappearing'.[62]

The most serious consequence of this educational ideal was that intellectual talents were attracted to fields of work and careers which did not, at least not directly, stimulate the English industrial development.[63] As was emphasised above, the higher technical education was mainly the responsibility of the universities, while Colleges of Advanced Technology – Civic Universities – held no more than a limited share in this responsibility.[64] The result partly of this state of affairs, but probably most of all of the 'markedly inferior status' of higher technical education,[65] was that the students at the universities preferred pure science – in case they did not study the humanities, law, etc. – to technology/engineering.[66]

Even around the middle of the twentieth century, it was generally held that the arts and pure sciences received 'an unfairly large proportion of the really able men';[67] an investigation at Oxford University in the early 1960s showed that English boys began studies in higher technology only if they had failed to secure a place at the faculty of pure science. According to the investigation, this meant that 'applied, as opposed to pure, science (was) getting the "rejects" . . .' Besides, it was emphasised that England deviated from the usual picture in Europe: 'To an extent unusual in Europe boys from the top grades of our science sixth cluster in the faculties of pure sciences, while engineering and dip. tech. courses draw on the lower grades'.[68]

The result of this investigation – which was considered to reflect the general opinion – was probably also an adequate general description of the English situation with regard to earlier decades of the twentieth century, and even more so in respect to the nineteenth century.

The reason for the passive role played by various governments in matters of organised and efficient technical education – as we recall, the first Technical Instruction Act was not passed until 1889 – is to be found in the social attitudes of a distinguished class society, too.

In connection with his discussion on the English neglect of professional training to promote innovation, Allen has expressed the point in the following manner: 'The chief blame must be assigned to what we now call "the establishment": political leaders, the Civil Service and those who sat on the boards of the chief industrial companies. They themselves were, of course, the victims of anachronistic institutions – the English class system and the educational arrangements associated with it.'[69]

The following conclusions can be drawn from our discussion on the factors explaining the English development.

The reason for the slow development of the higher technical education in England is partly to be found in the limited demand for these engineers on the part of industry, and in the limited career openings for engineers. This was, in its turn, connected with England's industrial successes during the earlier phases of the industrial revolution, and consequently also with the other conditions prevailing in the English society that have been discussed here: the development of the English engineering profession and the low social standing, especially among the upper and middle classes, of an education and career in engineering. For years, as we have seen, there was no united national engineering organisation serving the interests of the engineers; the consequences, not least with regard to the important supervision of the technical qualifications of engineers, were unfortunate. Besides, school talents preferred an education in the humanities, law, etc. and the pure sciences. From a social point of view, technology or engineering was considered an unsatisfactory choice. The failure of governments to promote an organised and state-supported technical education, not only at a high level but at all levels, must also be regarded against the background of social factors established in a long-standing class society.

Notes

1. P. Mathias, *The First Industrial Nation. An Economic History of Britain 1700-1914*, London 1969, p. 422.

2. W. Reid, *Memoirs and Correspondence of Lord Playfair* (1899), p. 110, after D.H. Thomas, *The Development of Technical Education in England 1851-1889, with special Reference to Economic Factors*, University of London 1940 (unpublished), p. 2. See also Argles, 1964, p. 13.

3. Second Report of the Royal Commissioners for the Exhibition of 1851 (1852), p. 14, quoted by Thomas, 1940, p. 35. The total number of exhibitors was 13 937; 6 861 came from the United Kingdom and 7 076 from thirty foreign countries. The United Kingdom won 78 of the 164 Council Medals that were awarded. British machinery 'swept the board' with 52 Medals, while France divided some 54 Medals fairly equally among her machinery, manufactures and raw materials. See Y. Ffrench, *The Great Exhibition: 1851*, London 1950. pp. 276-7.

4. See Musgrave, 1967, p. 36.

5. J. Scott Russel, *Systematic Technical Education for the English People*, London 1869, p. 8.

6. Ibid., p. 86.

7. Lyon Playfair in Scott Russel, 1869, p. 94.

8. In his *The Organisation of Science in England. A Retrospect*, London 1957, D.S.L. Cardwell writes: 'As a result there was alarm, near-panic, and a movement was set on foot that was, in all respects, more farreaching than any so far discussed – the Technical Instruction movement' (p. 84).

9. Two reports should be mentioned here: the Report of the Select Committee on the Provisions for giving Instruction in Theoretical and Applied Science to the Industrial Classes (1868), 'the Samuelson Commission' and the Reports of the Royal Commission on Scientific Instruction and the Advancement of Science (1871-75), 'the Devonshire Commission'.

10. According to Cardwell, very few of the recommendations of the Devonshire Commission were implemented (1957, p. 97).

11. Argles, 1964, p. 23.

12. Ibid., p. 53.

13. Cotgrove, 1958, p. 68.

14. Ibid., p. 15.

15. Cardwell, 1957, p. 127. Until the Technical Instruction Act of 1889, the subject of technical education was, according to Roderick and Stephens, 'being more or less experimented with'. See Gordon W. Roderick and Michael Stephens, *Education and Industry in the Nineteenth Century*, London 1978, p. 74. Among the important factors in this development, we find the so-called Polytechnic movement and the investigations on the system of technical education in other

countries that were made by the Royal Commission on Technical Instruction (1881-84). The Royal Commission on Depression of Trade and Industry (1886) should be mentioned too.

16. Argles, 1964, p. 39.

17. Landes, 1972, p. 344. The reluctance of the London capital market to invest in British industry has been considered wholly justified in this context 'for the rise in the amount of capital per head invested in industry after 1900 was not accompanied by any increase in productivity'. See Allen, 1979, p. 33.

18. Cardwell, 1957, p. 96.

19. Argles, 1964, p. 20.

20. Cardwell, 1957, p. 89.

21. Argles, 1964, p. 53.

22. See J.F. Baker, 'Engineering Education at Cambridge' in *Institution of Mechanical Engineers. Proceedings*, London 1957, pp. 991-2.

23. Here, we may point at University College, London – from the 1870s 'one of the chief places of engineering education in the country' – and Owens College, Manchester; 'one of the first in the field of applied science' (Argles, 1964, pp. 49-50), with a Chair in engineering in 1868.

According to E. Ashby, the foundation of the University of London in 1826 – of which University College, with a Chair in engineering in 1841, became a part – meant that 'the scientific revolution at last took root in English higher education'. See his *Technology and the Academics*, London 1963, p. 29. See also G. Haines, *German Influence Upon English Education and Science, 1800-1866*, New London, Connecticut, pp. 14, 17.

24. Argles, 1964, p. 45.

25. Cardwell, 1957, p. 156.

26. Roderick and Stephens, 1978, p. 107.

27. Argles, 1964, p. 72, after *Natural Science in Education*, HMSO, 1918.

28. E.J. Hobsbawm, *Industry and Empire*, Penguin Books 1970, p. 182.

29. It was opened in 1881 as the Normal School of Science.

30. Argles, 1964, p. 81.

31. This has also been emphasised by, for example, Cardwell, 1957, p. 156. See also S.B. Saul, *The Myth of the Great Depression 1873-1896*, London 1968, pp. 47-8, and Landes, 1972, p. 340.

32. M. Sanderson, *The Universities and British Industry 1850-1970*, London 1972, pp. 13-14.

33. Ibid.

34. Ibid., p. 60. In 1907, Oxford also established a Chair and a department in engineering.

35. Ibid., p. 119.

36. For instance, Cardwell, discussing the period 1888-1900, draws the following conclusion: 'education provided the best opportunity for the highly trained science graduate. Applied research, either in industry or under government could not offer anything like the same number of posts, and it is very unlikely that industrial technology could offer any comparable opportunities to the pure science graduate'. (1957, p; 144.) Similar views are expressed by Roderick and Stephens (1978, p. 147).

37. Besides, concerning the technical institutions of an English type that were created – the civic universities – they did not achieve university charters and status until the mid-1920s. See Argles, 1964, p. 71.

38. Similar views on the consequences of the English engineering education have been expressed by J.E. Gerstl and S.P. Hutton in their *Engineers: The Anatomy of a Profession. A Study of Mechanical Engineers in Britain*, London 1966, p. 7.

39. Landes, 1972, p. 344.

40. Ibid., p. 346.

41. Sanderson, 1972, p. 114. The other defects concern the failure to cover certain areas of industrial study, the limitations regarding the quantity of their student output, and the relative lack of support from industry.

42. Cotgrove, 1958, p. 187. See also Chapter 1, note 4, in this study.

43. Sanderson, 1972, p. 118.

44. Roderick and Stephens, 1978, p. 106.

45. Erickson, 1959, pp. 30-40.

46. Cotgrove, 1958, p. 187. See also, for example, Cardwell, 1957, pp. 182-9. It is notable that Roderick and Stephens seem to hold a different opinion on this point. They write: 'It is clear that industrial leaders who had lacked a scientific education were aware of their deficiences and were anxious to ensure that their own sons benefited from such an education. This concern often extended to their workmen too' and 'Leading industrialists were equally aware of the importance of pure science to industrial success.' (1978, p. 119.) Although there were probably many examples of such industrial leaders in England, the *general* pattern tells a different story. See also Chapter 1, note 12, in this study.

47. See Chapter 1, note 1.

48. However, it should be noted that during the eighteenth century, before the development of the steam engine and the growth of precision engineering, it is difficult to differentiate between, for example, mechanical and civil engineering. See S. Pollard, *The Genesis of Modern Management. A Study on the Industrial Revolution in Great Britain*, London 1965, p. 131 ff.

49. Reader, 1966, p. 122. In another context, however, Reader asserts that 'By 1860, or thereabouts, the elements of professional standing were tolerably clear' (p. 71).

50. Ibid., p. 149.

51. Ibid., p. 164.

52. L.T.C. Rolt, 1967, pp. 22, 63.

53. H.B. Thomson, *The Choice of a Profession* (1857), quoted by Reader, 1966, p. 70. In England, for an occupation to be considered as a profession, it had to be accepted as such among 'gentlemen', a concept which was peculiarly English. See Emmerson, 1973, p. 247, as well as note 58 below.

54. Not until the beginning of the 1960s did thirteen separate engineering organisations consider it expedient to create 'an avenue for their intercommunication and action on common concerns'. An Engineering Institutions Joint Council was created (1962), and a Royal Charter for the council – now called the Council of Engineering Institutions – was applied for. The Charter was granted in 1965. See Emmerson, 1973, pp. 263-4.

55. Reader, 1966, p. 117.

56. See, for example, Cotgrove, 1958, p. 155.

57. Coleman, 1973, p. 96.

58. Ibid., p. 97. It should be emphasised that the discussion deals with the English society. The Scottish society was free from gentlemanly connotations. See, for example, Emmerson, 1973, p. 251. Discussing the social origins of the British entrepreneurs, Mathias holds that they did not form a *class* but a *type* of person (1969, p. 156).

59. Ibid. This is what has been called 'the Buddenbrook syndrome'. However, summing up the discussion and the extant evidence on this point, P.L. Payne has emphasised that 'The fact is that the third generation argument remains unproven, and will remain so until more data are available on the longevity of firms and the location of effective internal control.' See his *British Entrepreneurship in the Nineteenth Century*, London 1974, p. 27. (Also pointed at in 'Industrial

Entrepreneurship and Management in Great Britain' in *Cambridge Economic History of Europe*, Vol. VII:1 (1978), pp. 202-3.)

60. Coleman, 1973, p. 101. Here, Coleman also refers to A.E. Musson and E. Robinson, *Science and Technology in the Industrial Revolution*, Manchester 1969, holding that during a certain time at the end of the seventeenth century, as well as 'in certain walks of life' during the eighteenth century, 'the first fine enthusiasm of the scientific revolution had diffused ideas of the new science and brought interests in its practical uses amongst all manner of men'.

61. See Coleman, 1973, pp. 97-101. Several definitions of the concept 'gentleman' are given in Coleman's article.

62. Ibid., p. 102. Ashby has also emphasised how 'The very stratification of English society helped to keep science isolated from its application' (1963, p. 51).

63. See below. See also Coleman's hypothetical reasoning (1973), p. 109 ff).

64. Ashby, 1963, p. 60.

65. See, for example, Argles, 1964, p. 138. Here, too, the view presented in this discussion – a view which deviates from Sanderson's – should be called to mind.

66. The fact that pure science was long held in higher esteem than applied science has been emphasised by, among others, R.S. Edwards and H. Townsend. See their *Business Enterprise. Its Growth and Organisation*, London 1958, p. 556.

67. Argles, 1964, p. 99.

68. *Technology and the Sixth Form Boy*, Oxford University, Department of Education (1963), quoted by Argles, 1964, p. 120.

69. Allen, 1979, p. 48. See also Roderick and Stephens, 1978, p. 167ff.

4 SUMMARY AND CONCLUSIONS

The process of economic growth and industrial performance has been studied by economists and economic historians for several decades. During the post-war period the interest in these studies increased substantially and new insights into the growth process were gained, especially since the residual or 'technic' factor was 'discovered'. Thus, it was established that a country's economic growth was to a large extent explained by factors other than the inputs of physical capital and labour only. In this context, investments in education, i.e. 'human capital', and technological change are of paramount importance.

In the entire process of growth and performance, the engineer is an important factor, especially at the innovatory stage, when basic research and new inventions are put to economic use. A thorough theoretical and practical technical education is necessary and has been so at least from those nineteenth-century years when the science-based industries generally assumed a leading position in the industrial sector of the economies.

However, although matters such as business organisation and costs on Research and Development (R & D) have been scrutinised for various countries, studies applying the results of this research to industry are relatively scarce; and although the quality of education in various countries has been discussed — for the twentieth as well as the nineteenth century — there has been no explicit discussion concerning the extent of the technical education system in general on a quantitative and comparative basis, let alone any investigation into a specific type of technical education. One reason for this is seen in the countries' unsatisfactory national statistics (see Chapter 1) and consequently our limited knowledge of the number of qualified and professionally active engineers during different periods of time.

The purpose of this study has been to increase our knowledge concerning the highly qualified engineers, i.e. those with an education from schools which were already, or were later during the nineteenth century converted into, technical universities. Calculations and estimations of their number and choice of careers make it possible to present a discussion from partly new angles of approach concerning the reason for and content of an institutionalised technical education, the demand for highly qualified engineers, the social status of an education and career in engineering, etc.

Although it is impossible to measure the importance of a specific type of engineer in the growth process, especially at a highly aggrega-ted level, the similarities and deviations found between the four coun-tries indicate, and support, the contention that highly qualified engineers played an essential part in the process of industrial growth and performance.

This study can be summarised in the following terms:

(1) An institutionalised type of higher technical education was introduced on the Continent at the end of the eighteenth century and the beginning of the nineteenth. France was the pioneering country, and although there were forerunners, it is in connection with the establishment of the École Polytechnique and its specialisation schools – especially the École des Mines and the École des Ponts et Chaussées – that it is relevant to speak of a formal higher technical education. The French system was important to the development in the German-speaking states and influenced the creation of technical schools of a new type, the Gewerbeinstitut. These schools, which were founded from the 1820s onwards, later developed into the Technische Hoch-schulen. Most important in the German development were the schools in Karlsruhe and Berlin, Karlsruhe because of the principle of the Fachschule (specialisation within the school; 1832) and later – from the 1860s – because of its vital part in the initiative and effort to bestow an academic character on these technical schools.

Quantitatively as well as qualitatively, the importance of the Berlin school is connected with its localisation and prestige in the united Germany. It also played the leading role during the latter decades of the nineteenth century in the long-drawn-out struggle for obtaining the same status as the classical universities.

A link in the general development of the higher technical education was the Zurich Technische Hochschule in the 1850s and 1860s. This school was also founded on the model of Karlsruhe, but in order to counterbalance too narrow an educational specialisation, lectures in, for example, humanistic and political disciplines were given.

These ideas found their way to the German technical schools, probably to a large extent *via* the technical university in Berlin. Thus, one can also point at notable personal links between the schools in Karlsruhe, Zurich and Berlin.

Summing up the most important chains in the development of the institutionalised higher technical education in Europe in the nine-teenth and early twentieth century, the technical schools situated in the following cities should be emphasised: Paris, Karlsruhe, Zurich and Berlin. However, Prague and Vienna were also of initial importance in the early-nineteenth-century development, and the first Swedish technical school of the new type had statutes that were originally very similar to the Vienna statutes. In the succeeding development of the two Swedish higher technical schools, the organisation and education were generally based on the German model.

The explicit purpose of the institutionalised technical education in France, Germany and Sweden was to improve the national industry. Normally, the schools were founded by the government, often together with private individuals, in order to provide the public and private sector of the economies with qualified engineers. It is notable, however, that the prestigious French École Centrale des Arts et Manufactures was founded as a result of private initiatives (1829), but the school was later (in 1857) taken over by the state. Thus, it was realised that the means to raise the national industry was a thorough (according to the standards of the time) technical education, which should include both theoretical and practical components. The latest knowledge gained in the field of science was considered an important element in the training. Consequently, 'scientific' principles in a technical education based on mathematics were introduced at an early stage. The only exception from this pattern among the schools studied here was the technical university in Stockholm – an unfortunate exception, in the view of the contemporary teaching staffs at the school – where the introduction of theoretical educational components was deferred until the middle of the nineteenth century.

(2) A substantial increase in the higher technical education occurred during the nineteenth century, and in the case of Germany the total number of students at the higher technical schools increased from around 1 000 at the middle of the nineteenth century to about 15 000 at the turn of the century. The annual number of graduates (flow) increased accordingly, and our calculations show an increase in the total number of highly qualified engineers (stock) in Germany from a mere 3 500 in the middle of the nineteenth century to more than 30 000 in the late 1880s, the number going up to and exceeding 60 000 at the time of World War I.

The French development presents a different picture. Thus, the calculations show a fourfold increase in the stock of qualified engineers during the first half of the nineteenth century, and by the middle of the nineteenth century the total number of these engineers in France was almost twice that of Germany. However, the further growth in France was lower than in Germany; and from the 1870s onwards, the number of engineers with this type of technical education was lower than in Germany in absolute terms. In the latter half of the 1880s it was two-thirds of the German number, a proportion that was unchanged at the time of the War. The stock of highly qualified engineers in France at that time thus amounted to approximately 40 000.

As a proportion of the economically active male population, the density of qualified engineers in mid-nineteenth century France was greater than in Germany, but below one per thousand. However, from the 1880s onwards there are no significant differences between the countries, and at the time of World War I there were three highly qualified engineers per thousand of the economically active male population in France as well as in Germany. The total number of qualified engineers in Sweden grew substantially, too, and in 1914 their total number amounted to 3 500. However, as a proportion of the economically active male population they only made up two per thousand at the time of the War, a figure that was reached in Germany and France in the 1880s and 1890s.

(3) Engineers with an advanced technical education were supposed to hold leading positions in the industrial activities and in the work on improving the technical methods. Qualified engineers in the private sector of the economy, with an education in mechanical engineering and professionally active in production and design, were usually considered to be essential to this process.

Against this background, it is notable that the aggregate figures on the stock of engineers at various times conceal the fact that in the case of France, a comparatively small number was active within industry, particularly in the private sector. This was actually the reason why the École Centrale des Arts et Manufactures was founded, as the demand of the private sector for qualified engineers was not met by the École Polytechnique and its specialisation schools. For example, only 10-30 per cent of the École des Mines (Paris) graduates were active within the private industry from 1830 up to World War I, and less than 10 per cent of the École Polytechnique graduates during 1870-1914. While almost 80 per cent of the polytechniciens during this period chose a purely military career, two thirds of those who joined the Corps Civil were in civil engineering.

Thus only a small proportion was active within production and design. It should be emphasised though, that École Polytechnique graduates as well as graduates from, for example, École des Mines and

École Centrale were active within all types of French industries, where they rapidly attained leading positions. However, although the Grandes Écoles engineers in production and design were comparatively few in number, they were probably to a large extent responsible for the pioneering tradition of French industry.

With regard to the supply of engineers to the private industry, the German case was completely different, a simple conclusion quickly obtained on the basis of the large number of graduated qualified engineers. Furthermore, the information available from certain studies can be seen to suggest that the nineteenth-century German labour market for the highly qualified engineers was balanced. It is a fact that a dominating proportion of the German industry, irrespectively of company size, possessed technically qualified engineers in management and leading positions towards the end of the nineteenth century and during the early twentieth century.

In the case of Sweden, qualified engineers went into all fields of occupation, but especially industry, and a dominating proportion to the private sector. The demand for them was great, and to a large extent they reached leading positions rather quickly.

(4) Like most students in higher education in nineteenth-century Europe, those in higher technical education in France, Germany and Sweden were usually drawn from the middle class, but a notably large and growing proportion came from the upper classes, too. From the very beginning of the nineteenth century, France appears to have had a slightly larger number of upper-class students in the technical field than the other countries. However, at the turn of the century in Sweden, for example, almost 95 per cent of the total number of students at the technical universities belonged to the upper and middle social classes, the highest group contributing almost 50 per cent of the total number.

Around 1800, France was the only country in the world where the engineering activity could be considered a learned profession, mainly as a result of the standing of the higher technical education in the country. In the middle of the century, France also obtained its national organisation of non-military engineers – *la Société des Ingenieurs Civils* – out of the student organisation of École Centrale.

In the German states there was a rapid development towards professional status in connection with the founding of the Gewerbe institutes and the associations formed by former students at these schools. In the 1850s the national *Verein Deutscher Ingenieure* was established with a view to furthering the professional interests of the engineers.

The Swedish Association of Engineers and Architects – *Svenska Teknologföreningen, STF* – developed in a similar way from the middle of the nineteenth century out of a student organisation at the Stockholm technical university. In the 1880s the STF became a national organisation, and amalgamation in the early 1890s with *Ingenjörsföreningen* in Stockholm, founded in 1865 – which included the best-known Swedish technicians – pushed the professional interests of engineers in the society further forward.

The status of higher technical education in Germany was growing during the nineteenth century in connection with the rapidly increasing interest in science and *Technik*. At the turn of the century, the technical universities also formally attained the same status as the classical university through the right to confer doctor's degrees. The Swedish development was, with a certain lag, similar to the German in this respect too, but although the right to confer doctor's degrees had been on the Stockholm school agenda for many years – the German development acting as a source of inspiration – this right was deferred until the 1920s.

In France, however, the engineering activity as such does not generally seem to have had a high status. This may seem somewhat paradoxical, considering that entrance requirements to French engineering schools have always been high. In comparison with the situation in Sweden, for example, the competition for a place at the most prominent French engineering schools was – and is – much harder.

The assertion made here concerning the engineering activity is partly based on the information supplied in the secondary literature and partly on information regarding the choice of careers of the Grandes Écoles students. I assume that there is a positive correlation between the choice of career and status in a particular occupation, although economic motives, for example, naturally play a role too. A connection between growth and fluctuations in the business cycle and the chances of securing a place at these schools can also be discerned, but in France it is less pronounced than in Sweden, for instance.

The choice of occupational fields by the Grandes Écoles students showed that on the whole only a small number of the students were active within engineering in any real sense. Thus it seems reasonable to conclude that the status and prestige which goes with a degree from one of the leading Grandes Écoles was and has always been high, whereas the engineering occupation as such has generally, but above all in mechanical engineering, had a lower status in the French society. A career in public service, particularly a military one, or the kind one might expect on joining the Corps Civil as a civil engineer, was more attractive. In general terms, it seems reasonable to speak of social differences in this respect between France on the one hand and Germany and Sweden on the other.

(5) England is the exception to the other countries. With respect to the institutional aspects of higher technical education, that is, the number and qualifications of these engineers, the demand for them, and their career patterns, etc., England generally differs from France, Germany and Sweden, although there are similarities to France concerning the social aspects of engineering occupations.

In England, the inauguration of a higher technical education came late in the day – around the middle of the nineteenth century – and its proportions were modest according to international standards. At the time of World War I the total number of full-time students in science and technology at universities, colleges and technical institutes was only about 3 000. Forty five per cent of these were studying at the technical institutes, which offered a training at a somewhat lower technical level, while a large proportion of the other formally more qualified students was studying at the classical universities, preferably in the pure sciences – not in engineering. The density of engineers with qualifications for membership in the leading engineering organisations was also significantly lower than in the other countries.

The government did not take much of an interest in such higher technical education as was supplied by institutions outside the academic sphere, institutions supposed to have been in close contact with the industry. Still, voices arguing in favour of the founding in England of technical universities of the Continental type were heard from the first. Despite several public investigations into the matter, however, little happened and not until around 1890 did the first Technical Instruction Act come into being. When the Imperial College of Science and Technology was finally founded in the early twentieth century out of three London higher technical institutions from the 1840s, 1850s and 1880s – the Royal College of Chemistry, the Royal School of Mines, the Royal College of Science – England received a concentrated institution of importance for higher technical education.

Looking into the reasons for the slow, and deviating, development in England I have emphasised three factors and/or indicators. Firstly, the low demand for highly qualified engineers and the engineer's limited chances of entering a career leading to a top post within business. This was also reflected in the difficulties of recruiting students to those few higher technical institutions that did exist.

Secondly, the unsatisfactory development of the engineering profession in the country – for example, the element of snobbery in that civil engineers looked down on mechanical engineers – combined with the lack of a united national organisation furthering the interests of engineers from economical and social points of view. In this context, the technical qualifications of engineers is an important factor, along with the social standing of an education and occupation in engineering.

Thirdly, the fact that engineering and/or technology was considered a socially unsatisfactory choice of career. This meant that school talents preferred an education in law, the humanistic disciplines, etc., and eventually the pure sciences. Studies in technology were generally their last choice. Thus, the most capable students, at least from a theoretical point of view, did not choose a career in engineering.

The passive role of government in issues concerning technical education – not only with regard to higher technical education, but at all levels – must also be seen against the background of cabinet members' own social and educational background, i.e. as a result of a prosperous class society, not adjusted to industrial development where application of the latest knowledge in science had become an important element.

Comparative studies of the type presented here are always associated with certain difficulties. Along with the conclusions I draw from the study, I want to indicate two of these: firstly, we may not in fact be comparing equivalent entities, i.e. in this case the highly qualified engineers educated in France, Germany and Sweden might be different as a result of, among other things, the organisation and contents of the kind of education they received. However, as the education in these countries fits into the general pattern of higher technical education, this aspect may be left out of account. It would seem, though, that the German education, which was imitated in Sweden, was preferable considering the needs of industry from the latter half of the nineteenth century. But a school such as the French École Centrale des Arts et Manufactures probably corresponded well to the highest demands of industry from the start, even when an international comparison is made. On the other hand, in the case of England (as has been emphasised above) there are several reasons to assume that the quality of the higher technical education, as well as the quality of the engineers, was generally lower than in the other countries.

Secondly, a highly aggregated quantitative study of the well-qualified engineers means that we overlook the special characteristics of the single country with regard to, for example, industrial structures. However, it is clear that at a stage in the industrial process when the importance of the science-based industries was growing into subsequent domination, the application of the latest science to industry became a necessity in an industrialised world of hardening competition.

A technical education based on the principles of 'learning by doing' and 'trial and error' was no longer enough. This was the case with France, Germany and Sweden but also with England, generally and irrespective of certain deviations in industrial structures (see below). I have here accepted the contemporary – and, looking back, self-evident – view concerning the need for qualified engineers within industry, especially in leading functions. The availability of a large number of these engineers was considered important for the industrial performance of the countries involved.

In an indicative way, this study has shown the relevance of the above view. When engineers with lower formal qualifications are also taken into account, the main result of the study is strengthened; of course, the technical education of man-power at all levels must be considered in a nation's educational system.

Germany stands out here as the successful case in point. This also goes for Sweden, although the number of highly qualified engineers as a proportion of the economically active male population was somewhat lower in that country than in Germany, as well as in France. But in the case of France the number of highly qualified engineers active within the industry in production and design was more limited; in addition, there were too few engineers and technicians for middle-range positions. In England there was generally an 'unmet need' for qualified engineers (see Chapter 1:2), but especially for those with a highly qualified technical education, i.e. demand and supply were both too small for the nation's industrial needs.

From the latter half of the nineteenth century, the structure of British industry in comparison with the other countries – especially Germany – was superannuated, and the passage of time exacerbated the problem. Besides, the growth of the British industry was – as was pointed out in the introductory chapter – lower than in Germany, but not – at least not in a long-term perspective – lower than in France.

From a technological point of view, however, there were general differences between Britain and the Continent. The relative British decline has to a large extent been discussed in the context of Britain's being the first nation in the world to become industrialised and her consequent inability and unwillingness to adjust the economy to changed industrial-technological conditions.[1] There are, as I have

emphasised, important sociological and cultural reasons for support-
ing this view. However, from a purely economical and theoretical point
of view, the idea that being the first in the process of industrialisation
necessarily constitutes a disadvantage is hardly a convincing argument.
It is thus equally possible to argue for the advantage of 'forwardness'
in a period of accelerated technological change.[2] The amount of
resources in the 'old' nation is larger, and hence also the economic
possibilities of making new investments. Latecomers, of course, possess
comparative advantages in that they know the older methods of
production and materials and can consequently avoid mistakes, cut
corners, etc. But there is no theoretical argument from a technological
point of view to the effect that the firstcomer should be overtaken by
the latercomers. Any judgement as to whether a nation's being early
or late in this respect constitutes an advantage or a drawback, must be
the result of an economic-historical analysis. However, the notion that
Britain was, during the nineteenth century, overtaken by its Continental – and American – competitors in industrial sectors where science
and its application were of paramount importance, seems to be true.
The unwillingness of the British chemical industry to change over from
the Leblanc method to the Solvay process in making soda – a process
that was introduced on the Continent in the 1870s – is considered a
case in point here.

In this context, however, the British iron and steel industry and its
relative decline from the latter decades of the century onwards have
come in for a larger share of study and discussion. This has also had a
decisive effect on the general opinion concerning the British economy
of the period, according to which the decline was heavier than necessary. The responsibility has been attributed to factors such as an
inefficient capital market, slow growth in foreign and internal markets,
and a weakness in the British entrepreneurship caused by inefficiency
in the educational system and a lack of technical competence in entrepreneurs and personnel.

Without denying the relative decline of the British iron and steel
industry, new economic historians have questioned this view of the
British entrepreneur. Instead, they have emphasised that the decline
was unavoidable and that it occurred with retained efficiency in production. Conclusions along this line, but in which the argument of
unavoidability was questioned, have also been put forward. In comparison with the German development, it was hence emphasised that the
British iron and steel industry was the base for different 'development
blocks' – in the terminology of Dahmén – and that a weakness in the

British structure from the aspect of growth was that the production of technologically modern machines did not attain the importance it could have had. This was not due to the British being too conservative and too traditional, but rather the opposite, that is, the British being too quick in adopting the new technology.[3]

To take a definite stand among the arguments concerning the relative merits of being 'first' in the industrialisation process, i.e. a stand that forms a kind of objective view, is very difficult. It boils down to a question of the relative weight of various factors, factors which are not easy — often impossible — to identify, isolate and measure.

At the base of such a discussion, though, there must be a question as to what choice of investments and industrial technologies was made. Normally, the various kinds of action taken here were the results of decisions made by individuals in management and leading positions in firms and companies. The 'right' decision always requires a thorough knowledge of the market and the industrial processes, from a technological point of view too. The long-term views are vital, i.e. the ability to see the development and growth potentials of an investment. For obvious reasons, the studies by new economic historians have overlooked this aspect and have had to concentrate on the efficiency of the short term.

Whether the long-term perspectives have been foreseen can only be established with the aid of hindsight, but here most studies indicate a general British failure — and there is no getting past the need for generalisations. Landes' view concerning the British industry from the 1870s onwards is still relevant: although there were bright spots in British industry, these cannot conceal 'the *generally* lustreless performance' of the industry. This general point was not only attributed to the older branches of industry; the performance of newer industries was also 'either spotty and sporadic' (electricity and electrical engineering) or 'embarrassingly weak' (chemicals). On the technological frontier the 'best practice was *generally* not so advanced' as, for example, in Germany. The worst symptom of Britain's industrial situation was 'the extent to which her entrepreneurship and technology were defensive'.[4]

In connection with the above discussion, Landes also pointed at two very important symptoms, namely the tendency for British innovations to be exploited more rapidly and effectively abroad, and the contribution to industrial technique made by people of foreign birth and training or by Britons who learned their trades abroad.

Behind all these indicators and symptoms we find the factor of technical qualification and its personalisation in the engineer, mostly in production and design. The British 'unmet need' for qualified engineers must − to take an expression from Saul in his discussion of the British preference for the practical rather than the theoretically-trained man − be seen as 'a legacy of an early start in industrialization . . .'.[5] In this respect it is relevant to talk of the pioneer's disadvantage − not because of any inherent necessity, but because earlier industrial successes had caused effects of a social and cultural kind, which at later stages in the industrial development worked against the introduction and acceptance of a necessary advanced technical education.

The low social esteem in which the English engineer was held, and his limited career chances, were also part of these effects. The Continental pattern of technical education in general, but especially at the highest level, the number of the engineers, their careers and positions, and the various technological and industrial performances of France, Germany and Sweden strongly indicate the importance of the qualified engineer in the process of growth in the nineteenth and early twentieth centuries.

Certainly, many things have changed in this respect during subsequent decades; but a student of the British economy cannot but observe that the situation still seems to persist to some extent in the Britain of the early 1980s.[6]

Notes

1. For an extensive discussion of the views on the British situation, see Saul, 1968, and Payne, 1974; 1978.

2. Expression from A. Shonfield in *Modern Capitalism. The Changing Balance of Public and Private Power*, London 1970, p. 59.

3. See C-A. Nilsson, *Järn och stål i svensk ekonomi 1885-1912. En marknadsstudie*, Lund 1972, chapter 2 and p. 148.

The case of Britain in the research is here discussed in a separate section.

4. D.S. Landes, 'Factor Costs and Demand: Determinants of Economic Growth. A Critique of Professor Habakkuk's Thesis', *Business History*, Vol. VII:1 (1965), pp. 25-6. Italics added.

5. Saul, 1968, p. 48. Playfair's words from the middle of the nineteenth century are worth remembering here: 'We still hold to mere experience as the sheet anchor of this country, forgetful that the moulds in which it was cast are of antique shape, and ignorant that new currents have swept away the sand which formerly held it fast, so that we are in imminent risk of being drifted ashore.' See L. Playfair, 1852, p. 7.

6. See 'Repairing Britain's engineers', *The Economist*, February 1981.

APPENDIX

Calculated Number of Engineers with Higher Technical Education —
France, Germany, Sweden. Annual Numbers

	France	Germany [b]	Sweden
1800	1 500	–	–
1810	2 590	–	–
1820	3 467	–	–
1830	4 416	–	..
1831	..	–	40
1832	..	–	65
1833	..	..	93
1834	..	..	129
1835	..	166	162
1836	..	361	184
1837	..	553	219
1838	..	735	251
1839	..	925	288
1840	5 805	1 108	334
1841	..	1 302	372
1842	..	1 505	409
1843	..	1 717	446
1844	..	1 940	484
1845	..	2 144	523
1846	..	2 343	555
1847	..	2 570	575
1848	..	2 815	594
1849	..	3 071	613
1850	6 687[a]	3 343	637
1851	6 881	3 630	653
1852	7 082	3 922	674
1853	7 291	4 249	683
1854	7 508	4 605	699
1855	7 733	4 935	733
1856	7 966	5 236	739
1857	8 207	5 547	766
1858	8 457	5 886	805
1859	8 716	6 274	810
1860	8 972	6 731	854
1861	9 237	7 292	885
1862	9 512	7 878	913
1863	9 797	8 473	944
1864	10 092	9 028	974

1865	10 397	9 550	999
1866	10 713	10 018	1 026
1867	11 040	10 485	1 043
1868	11 378	10 980	1 069
1869	11 727	11 411	1 106
1870	12 050	11 856	1 121
1871	12 385	12 423	1 146
1872	12 732	13 132	1 163
1873	13 092	13 657	1 174
1874	13 465	14 994	1 191
1875	13 851	16 400	1 217
1876	14 251	17 900	1 265
1877	14 665	19 502	1 293
1878	15 093	21 294	1 341
1879	15 536	22 970	1 378
1880	15 994	24 452	1 406
1881	16 468	25 754	1 429
1882	16 958	26 836	1 443
1883	17 464	27 726	1 455
1884	17 987	28 434	1 454
1885	18 527	29 062	1 457
1886	19 085	29 675	1 468
1887	19 661	30 265	1 483
1888	20 256	30 870	1 512
1889	20 870	31 495	1 548
1890	21 504	32 166	1 612
1891	22 155	32 948	1 647
1892	22 823	33 767	1 701
1893	23 508	34 773	1 754
1894	24 211	36 036	1 805
1895	24 932	37 467	1 866
1896	25 672	39 245	1 962
1897	26 431	41 257	2 017
1898	27 210	43 457	2 083
1899	28 009	45 921	2 157
1900	28 829	41 657	2 237
1901	29 672	44 153	2 312
1902	30 537	46 974	2 396
1903	31 424	50 144	2 481
1904	32 334	53 737	2 618
1905	33 268	50 178	2 711
1906	34 227	52 265	2 795
1907	35 210	54 294	2 894
1908	36 219	56 191	2 979
1909	37 254	58 018	3 069

1910	38 317	59 738	3 145
1911	39 407	61 296	3 216
1912	40 526	62 701	3 317
1913	41 674	64 109	3 427
1914	42 850	65 202	3 504

Notes:

[a] Calculations starting from the figure 1 500 in year 1800, gave a total of 6 491 at the end of 1850. This figure, approximated to 6 500, is the base for the calculations starting in 1850, which gives us a total number of 6 687 at the end of that year.

[b] Based on figures of total number of students at the schools in winter terms.

Comment: For sources and methods of calculation, see Chapter 2, notes 56 (France), 59 (Germany) and 62, 63 (Sweden).

BIBLIOGRAPHY

Official Statistics and Reports

(Archival materials pertaining to the schools have also been consulted. References to such material are found in the relevant sections.)

Undervisningsväsendet. Statens Allmänna Läroverk för Gossar. BiSOS (Stockholm) 1897

Historisk Statistik för Sverige, Part I, Population, Stockholm 1969

Betänkande och förslag angående den lägre tekniska undervisningen i riket, 21 November 1874. Printed in *Bihang till Riksdagens protokoll vid lagtima riskdagen i Stockholm år 1876*

Betänkande och förslag till utvidgning och omorganisation af Tekniska Högskolan, Stockholm 1891

Utlåtande och förslag till den lägre undervisningens ordnande. Committee of 1907. Printed in *Bihang till Riskdagens protokoll vid lagtima Riskdagen i Stockholm 1918*

Betänkande med undersökningar och förslag i anledning av tillströmningen till de intellektuella yrkena. SOU (Stockholm) 1935:52.

Betänkande med utredning och förslag angående den högre tekniska undervisningen. SOU (Stockholm) 1943:34

Second Report of the Royal Commissioners for the Exhibition of 1851 (1852)

Report of the Schools Inquiry Commissioners on Technical Education. Parliamentary Papers, 1867, Vol. 26

Report of the Select Committee on the Provisions for giving Instruction in Theoretical and Applied Science to the Industrial Classes (1868) — the Samuelson Commission

Reports of the Royal Commission on Scientific Instruction and the Advancement of Science (1871-75) — the Devonshire Commission.

The Royal Commission on Technical Instruction (1881-84)

The Royal Commission on Depression of Trade and Industry (1886)

Reports of the Royal Commissioners on University Education in London (1910-13)

Natural Science in Education. The Report of the Committee on the Position of Natural Science in the Educational System of Great Britain (1918)

Report of Committee on Industry and Trade (1927) — the Balfour Committee

Books, Pamphlets, Articles, etc.

Abrahamsson, B. *Militärer, makt och politik*, Stockholm, 1972

Adler, V. *Om det tekniska undervisningsväsendet i Sverige*, Stockholm, 1897

Aguillon, M. *Supplément à la Notice Historique sur L'École Nationale Supérieure des Mines*, Paris, 1899

Allen, G.C. *The British Disease*, London, 1979

Althin, T. *KTH 1912-62. Kungl. Tekniska Högskolan i Stockholm under 50 år*, Uppsala, 1970

André, D. *Indikatoren˙ des technischen Fortschrittes*, Göttingen, 1971

Ardagh, J. *The New French Revolution. A Social & Economic Survey of France 1945-1967*, London, 1968

Argles, M. *South Kensington to Robbins. An Account of English Technical and Scientific Education since 1851*, London, 1964

Artz, F.B. *The Development of Technical Education in France 1500-1850*, Cambridge, Mass., 1966

Ashby, E. *Technology and the Academics*, London, 1963

Babbage, Ch. *Reflections on the Decline of Science in England and on Some of its Causes*, London, 1830

Baker, J.F. 'Engineering Education at Cambridge', *Institution of Mechanical Engineers. Proceedings*, London, 1957

Baucher, E., Moore, A. *La Formation et le Recrutement de l'Ingénieur Civil des Mines 1817-1939*, 1973 (unpublished)

Bodman, G. (ed.), *Chalmers Tekniska Institut. Minnesskrift, 1829-1929*, Gothenburg, 1929

Bradley, M. 'Scientific Education for a New Society. The École Polytechnique 1795-1830', *History of Education*, Vol. 5:1, 1976

Callot, J.P. *Histoire de l'École Polytechnique*, Paris, 1958

Cardwell, D.S.L. *The Organisation of Science in England. A Retrospect*, London, 1957

Chronik der Königlichen Technischen Hochschulen zu Berlin 1884-1899, Berlin, 1899

Coignet, M.E. *Rapport présenté au Conseil de l'École Centrale. Au nom de la Commission spéciale chargée d'examiner la question dite 'De la Spécialisation'*, Paris, 1910

Coleman, D.C. 'Gentlemen and Players', *Economic History Review*, 2nd Series, Vol. XXVI:1, 1973

de Comberousse, Ch. *Histoire de l'École Centrale des Arts et Manufactures*, Paris, 1879

Cotgrove, S.F. *Technical Education and Social Change*, London, 1958

Dahn, P. *Studier rörande den studerande ungdomens geografiska och sociala härkomst*, Lund, 1936

Deane, Ph. *The First Industrial Revolution*, Cambridge, 1969

Die deutschen technischen Hochschulen. Ihre Gründung und geschichtliche Entwicklung, Munich, 1941

Depeaux, G. *La Société Française, 1789-1960*, Paris, 1964

l'École Centrale, Annuaire, 1976

École des Mines, Armines, Rapport d'Activité, 1972

E.N.S. des Mines de Paris. Cycle de Formation des Ingénieurs Civil, 1974

Les Écoles Nationales Supérieurs des Mines, Nr 15, Novembre 1961

Edwards, R.S., Townsend, H. *Business Enterprise. Its Growth and Organisation*, London, 1958

Eidgenössische Technische Hochschule 1855-1955. École Polytechnique Fédérale, Zurich, 1955

Emmerson, G.S. *Engineering Education: A Social History*, Newton Abbot, 1973

Erickson, Ch. *British Industrialists. Steel and Hosiery 1850-1950*, Cambridge, 1959

Eriksson, G. *Kartläggarna. Naturvetenskapernas tillväxt och tillämpningar i det industriella genombrottets Sverige 1870-1914*, Umeå, 1978

Ffrench, Y. *The Great Exhibition: 1851*, London, 1950

Fischer, W. *Der Staat und die Anfänge der Industrialisierung in Baden 1800-1850*, Band I, 'Die staatliche Gewerbepolitik', Berlin, 1962

—— 'Government Activity and Industrialization in Germany (1815-1870)' in W.W. Rostow (ed.), *The Economics of Take-Off into Sustained Growth*, London, 1963

Fohlen, C. 'The Industrial Revolution in France 1700-1914', *The Fontana Economic History of Europe*, Vol. IV:1, 1973

—— 'Entrepreneurship and Management in France in the Nineteenth Century', *The Cambridge Economic History of Europe*, Vol. VII:1, 1978

Gårdlund, T. *Industrialismens samhälle*, Stockholm, 1942

Gerschenkron, A. *Economic Backwardness in Historical Perspective*, New York, 1965

Gerst, J.E., Hutton, S.P. *Engineers: The Anatomy of a Profession. A Study of Mechanical Engineers in Britain*, London, 1966

Goldbeck, G. *Technik als geistige Bewegung in den Anfängen des deutschen Industriestaates*, Berlin, 1934

Grüner, G. *Die Entwicklung der höheren technischen Fachschulen im*

112 Bibliography

deutschen Sprachgebiet, Braunschweig, 1967

Guillet, L. *Cent Ans de la Vie de l'École Centrale des Arts et Manufactures, 1829-1929*, Paris, 1929

Guinchard, J. *Sweden. Historical and Statistical Handbook*, Vol. I, Stockholm, 1914

Hachette, M. *Correspondance sur l'École Imp. Pol.*, Paris, 1813

Haines, G. *German Influence Upon English Education and Science, 1800-1866*, New London, Connecticut, 1957

v. Handorff, F. 'Die Verwendung der Hochschulsabsolventen im Staatsdienst, in der städtischen Werken und Verwaltungen und in der Industrie', *Abhandlungen und Berichte über technisches Schulwesen*, Band IV, Berlin, 1912

Hardach, K.W. 'Some Remarks on German Historiography and its Understanding of the Industrial Revolution in Germany', *The Journal of European Economic History*, Vol. I, 1972

Henriques, P. *Skildringar ur Kungl. Tekniska Högskolans Historia*, Vol. I-II, Stockholm, 1917 and 1927

Hobsbawm, E.J. *Industry and Empire*, Penguin Books, 1970

Holmberger, G. *Svenska Teknologföreningen 1861-1911*, Stockholm, 1912

Ingénieurs Diplômés 58/59, Octobre 1974

Ingeniör Lexikon, Stockholm 1805

International Encyclopedia of Social Sciences, Vols. 5, 15, 1968

Jewkes, J. 'How much Science?', *The Economic Journal*, Vol. LXX, 1960

Karmarsch, K. *Die Polytechnische Schule zu Hannover*, Hannover, 1856

Kemp, T. *Industrialization in Nineteenth-Century Europe*, London, 1969

Kindleberger, Ch.P. *Economic Growth in France and Britain 1851-1950*, Cambridge, Mass., 1964

—— 'Germany's Overtaking of England 1806-1914', *Weltwirtschaftliches Archiv*, Bd 111:3, 1975

—— 'Technical Education and the French Entrepreneur', in Carter II, Ed.C., Foster, R., and Moody, J.N. (eds.) *Enterprise and Entrepreneurs in Nineteenth and Twentieth Century France*, Baltimore, 1976

—— *Economic Response. Comparative Studies in Trade, Finance and Growth*, Cambridge, Mass., and London, 1978

Klages, H., Hortleder, G. 'Gesellschaftsbild und soziales Selbstverständnis des Ingenieurs', *Schmollers Jahrbuch*, 85:6, 1965

Kocka, J. 'Entrepreneurs and Managers in German Industrialization',

The Cambridge Economic History of Europe, Vol. VII:1, 1978

Kuznets, S. *Modern Economic Growth. Rate, Structure and Spread*, London, 1966

Laffitte, P. *Les Écoles d'Ingénieurs en France. La documentation française* Nos 4045-4046-4047, 3 Decembre 1973

Landes, D.S. 'French Entrepreneurship and Industrial Growth in the Nineteenth Century', *The Journal of Economic History*, Vol. IX:1, 1949

—— 'French Business and the Businessman: A Social and Cultural Analysis' in Earle, E.M. (ed.), *Modern France. Problems of the Third and Fourth Republics*, Princeton, 1951

—— 'Factor Costs and Demand: Determinants of Economic Growth. A Critique of Professor Habakkuk's Thesis', *Business History*, Vol. VII:1, 1965

—— *The Unbound Prometheus. Technological Change and Industrial Development in Western Europe from 1750 to the Present*, Cambridge, 1972

Layton Jr, E.T. 'American Ideologies of Science and Engineering', *Technology and Culture*, Vol. 17:4, 1976

Levy-Leboyer, M. 'Le patronat français a-t-il été malthusien?' *Le Mouvement Social*, Nr 88, Juillet-Septembre, 1974

Lexis, W. *Das Unterrichtswesen im Deutschen Reich*, Vol. IV, Berlin, 1904

Lilley, S. 'Technological Progress and the Industrial Revolution 1700-1914', *The Fontana Economic History of Europe*, Vol. III, 1973

Ludwig, K-H. *Technik und Ingenieure im Dritten Reich*, Düsseldorf, 1974

Lundborg, T. 'En blick på tekniska högskolans historia', *Teknisk Tidskrift*, 1927

Lundgreen, P. *Techniker in Preussen während der frühen Industrialisierung. Ausbildung und Berufsfeld einer entstehenden sozialen Gruppe*, Berlin, 1975

Malmsten, K. 'Från krigsingenjör till bergsingenjör', *Daedalus, Tekniska Museets Årsbok*, Stockholm, 1942

Manegold, K-H. *Universität, Technische Hochschule und Industrie. Ein Beitrag zur Emanzipation der Technik im 19. Jahrhundert unter besonderer Berücksichtigung des Bestrebungen Felix Kleins*, Berlin, 1970

Mathias, P. *The First Industrial Nation. An Economic History of Britain 1700-1914*, London, 1969

McClellan, W. 'A Suggestion for the Engineering Profession', *The Trans-*

action of the American Institute of Electrical Engineers, 1913

Mercie, C. *Les Polytechniciens 1870-1930. Recrutement et Activités* (unpublished), 1972

Mitchell, B.R. *European Historical Statistics 1750-1970*, London, 1975

v. Mohl, R. *Die Polizeiwissenschaft nach den Grundsätzen des Rechtsstaates*, Tübingen, 1844

Musgrave, P.W. *Technical Change, the Labour Force and Education. A study of the British and German iron and steel industries 1860-1964*, Oxford, 1967

Musson, A.E., Robinson, E. *Science and Technology in the Industrial Revolution*, Manchester, 1969

Musson, A.E. (ed.), *Science, Technology and Economic Growth in the Eighteenth Century*, London, 1972

Nilsson, C-A. *Järn och stål i svensk ekonomi 1885-1912. En marknadsstudie*, Lund, 1972

Palmstedt, C., Schoultz, Ed.v. *Historisk öfversigt af Chalmerska stiftelsens och statens teknologiska läroanstalts tillkomst och undervisningsarbeten, m.m.*, Gothenburg, 1869

Payne, P.L. *British Entrepreneurship in the Nineteenth Century*, London, 1974

——— 'Industrial Entrepreneurship and Management in Great Britain', *Cambridge Economic History of Europe*, Vol. VII:1, 1978

Pfetsch, F.R. *Zur Entwicklung der Wissenschaftspolitik in Deutschland 1750-1914*, Berlin, 1974

Playfair, L. *Industrial Instruction on the Continent*, London, 1852

Pollard, S. *The Genesis of Modern Management. A Study on the Industrial Revolution in Great Britain*, London, 1965

Popitz, H., Bahrdt, H-P, *Technik und Sozialarbeit. Soziologische Untersuchung in der Huettenindustrie*, Tübingen, 1957

'Le Prix des Cadres', *L'Expansion*, Juin 1976

Reader, W.J. *Professional Men. The Rise of the Professional Classes in Nineteenth-Century England*, London, 1966

Reid, W. *Memoirs and Correspondence of Lord Playfair*, London, 1899

'Repairing Britain's Engineers', *The Economist*, February 1981

Repertoire de l'École Imperiale Polytechnique, Paris 1855 and 1867

Roderick, G.W., Stephens, M. *Education and Industry in the Nineteenth Century*, London, 1978

Rolt, L.T.C. *The Mechanicals. Progress of a Profession*, London, 1967

——— *Victorian Engineering*, The Penguin Press, Pelican Books, 1974

Rosenberg, N. *Technology and American Economic Growth*, New York, 1972

Runeby, N. *Teknikerna, vetenskapen och kulturen. Ingenjörsundervisning och ingenjörsorganisationer i 1870-talets Sverige*, Uppsala, 1976

Rürup, R. 'Historians and Modern Technology', *Technology and Culture*, Vol. 15:2, 1974

Russel, J. Scott *Systematic Technical Education for the English People*, London, 1869

Rystedt, C.G. 'Teknologernas verksamhet och öden', *Industritidningen Norden*, 1881

Sanderson, M. *The Universities and British Industry 1850-1970*, London, 1972

Saul, S.B. *The Myth of the Great Depression 1873-1896*, London, 1968

Schimank, H. *Der Ingenieur. Entwicklung eines Berufes bis Ende des 19. Jahrhunderts*, Cologne, 1961

Schmookler, J. *Invention and Economic Growth*, Cambridge, Mass., 1966

Schnabel, F. 'Die Anfänge des Hochschulewesens', in *Festschrift anlässlich des 100jährigen Bestehens der Technischen Hochschule Fridericiana zu Karlsruhe*, Karlsruhe, 1925

—— *Deutsche Geschichte im 19.Jahrhundert*, Bd III, Freiburg, 1954

Shonfield, A. *Modern Capitalism. The Changing Balance of Public and Private Power*, London, 1970

Schröder, F. *Die höheren technische Schulen nach ihrer Idee und Bedeutung*, Braunschweig, 1847

Statistik der Grossherzoglich Badischen Technischen Hochschule zu Karlsruhe. Weltausstellung zu Chicago 1893, Karlsruhe, 1893

Sylvan, P., Kuylenstierna, O. (eds.), *Minnesskrift med anledning av K. högre artilleriläroverkets och krigshögskolans å Marieberg samt Artilleri-och ingenjörshögskolans etthundraåriga tillvaro, 1818-1918*, Stockholm, 1918

Die Technische Hochschule zu Berlin 1799-1924. Festschrift, Berlin, 1925

Technology and the Sixth Form Boy, Oxford University, Department of Education, 1963

Thomas, D.H. *The Development of Technical Education in England 1851-1889, with Special Reference to Economic Factors* (unpublished), University of London, 1940

Thomson, H.B. *The Choice of a Profession*, London, 1857

Torstendahl, R. *Teknologins Nytta. Motiveringar för det svenska, tekniska utbildningsväsendets framväxt framförda av riksdagsmän och utbildningsadministratörer 1810-1870*, Uppsala, 1975

—— *Dispersion of Engineers in a Transitional Society. Swedish Tech-*

nicians 1860-1940, Uppsala, 1975

Treue, W. 'Das Verhältnis der Universitäten und Technischen Hochschulen zueinander und ihre Bedeutung für die Wirtschaft' in Lütge, F. (ed.), *Die wirtschaftliche Situation in Deutschland um die Wende vom 18. zum 19. Jahrhundert*, Stuttgart, 1964

Treue, W., Pönicke, H., Manegold, K-H. *Quellen zur Geschichte der industriellen Revolution*, Göttingen, 1966

Wallander, J. 'Ingenjörerna i studentbetygen och i verkligheten', *Teknisk Tidskrift*, 1944

Wallmark, L.J. *Om tekniska elementar-skolors inrättande i Sverige*, Stockholm, 1851

Wickenden, W.E. *A Comparative Study of Engineering Education in the United States and in Europe*, Lancaster, Pa., 1929

Wijkander, A. *Chalmerska Institutet 1829-1904*, Gothenburg, 1907

Zöller, E. *Die Universitäten und Technische Hochschulen*, Berlin, 1891

Zvorikine, A. 'Ideas of Technology. Technology and the Laws of its Development', *Technology and Culture*, Vol. 3:4, 1962

AUTHOR INDEX

Abrahamsson, B. 22
Adler, V. 57
Aguillon, M. 45
Allen, G.C. 21, 80, 89
Althin, T. 36
André, D. 48
Ardagh, J. 37
Argles, M. 16, 79, 80, 81, 82, 83, 88, 89
Artz, F.B. 21, 22-3, 30, 31, 32, 41, 44, 46, 60
Ashby, E. 81, 88

Babbage, Ch. 17
Bahrdt, H-P. 22
Baker, J.F. 81
Baucher, E. 46
Bodman, G., 29, 40, 49
Bradley, M. 31

Callot, J.P. 61
Cardwell, D.S.L. 34-5, 80, 81, 82, 83, 85
Coignet, M.E. 41
Coleman, D.C. 16, 88
de Comberousse, Ch. 31, 32, 41, 42, 43, 61
Cotgrove, S.F. 16, 80, 85, 86

Dahn, P. 59
Deane, Ph. 17
Depeaux, G. 46

Edwards, R.S. 89
Emmerson, G.S. 21 passim, 34, 86, 88
Erickson, Ch. 16, 85
Eriksson, G. 16, 57

Ffrench, Y. 79
Fischer, W. 41, 54
Fohlen, C. 32, 63

Gårdlund, T. 35
Gerschenkron, A. 16, 63
Gerstl, J.E. 84
Goldbeck, G. 56 passim
Grüner, G. 29, 34
Guillet, L. 61

Guinchard, J. 36

Hachette, M. 61
Haines, G. 81
v. Handorff, F. 48
Hardach, K.W. 29
Henriques, P. 29, 33, 36, 40, 49, 61
Hobsbawm, E.J. 81
Holmberger, G. 23, 24, 53
Hortleder, G. 55
Hutton, S.P. 84

Jewkes, J. 19

Karmarsch, K. 55
Kemp, T. 42-3, 54
Kindleberger, Ch.P. 16, 20, 61, 62 passim
Klages, H. 55
Kocka, J. 54
Kuylenstierna, O. 35, 40
Kuznets, S. 13

Laffitte, P. 61
Landes, D.S. 15, 29, 47-8, 55, 61 passim, 63 passim, 80, 82, 84-5, 104
Layton Jr, E.T. 22
Levy-Leboyer, M. 37, 42, 43, 44, 47, 60
Lexis, W. 34
Lilley, S. 16
Ludwig, K-H. 20, 38, 56
Lundborg, T. 36
Lundgreen, P. 47, 54-5

McClellan, W. 22
Malmsten, K. 21
Manegold, K-H. 23, 33, 34, 35, 38, 55, 56
Mathias, P. 79, 88
Mercie, C. 45
Mitchell, B.R. 13, 14, 38
v. Mohl, R. 23
Moore, A. 46
Musgrave, P.W. 16, 79
Musson, A.E. 17, 88

117